Applications of Data Science to Molecules and Materials 2022

AA009260

Topical Conference at the 2022 AIChE Annual Meeting

Phoenix, Arizona, USA
13-18 November 2022

ISBN: 978-1-7138-7918-3

Printed from e-media with permission by:

Curran Associates, Inc.
57 Morehouse Lane
Red Hook, NY 12571

Printed with permission by Curran Associates, Inc. (2023)

For permission requests, please contact AIChE
at the address below.

AIChE
120 Wall Street, FL 23
New York, NY 10005-4020

Phone: (800) 242-4363
Fax: (203) 775-5177

www.aiche.org

Additional copies of this publication are available from:

Curran Associates, Inc.
57 Morehouse Lane
Red Hook, NY 12571 USA
Phone: 845-758-0400
Fax: 845-758-2633
Email: curran@proceedings.com
Web: www.proceedings.com

TABLE OF CONTENTS

APPLICATIONS OF DATA SCIENCE IN CATALYSIS AND REACTION ENGINEERING

APPLICATIONS OF DATA SCIENCE IN MOLECULAR SCIENCES I

APPLICATIONS OF DATA SCIENCE TO HIGH THROUGHPUT EXPERIMENTATION

INNOVATIONS IN METHODS OF DATA SCIENCE

APPLICATIONS OF DATA SCIENCE IN MOLECULAR SCIENCES II

283a Predicting the Performance of Mesophase Formation and Properties of Mesophase Pitch Based on Experimental Investigation and Machine Learning

Wenjia Wang, Madison Cooley, Karissa Jolley, Robert M. Kirby, Eric Eddings

Producing value-added mesophase pitch (MP) as the intermediate of high-performance carbon materials (carbon fibers, carbon foams, and graphite electrodes) via thermal polymerization is a promising process to utilize low-rank carbonaceous resources and waste-like bituminous coals, coal tar pitch, asphaltene, biomass, and waste plastics [1]. In this study, an orthogonal experimental design was applied to determine the effects of composition and operational conditions on the production of MP. However, effective optimizing operational conditions for desired product production through experiments can be energy- and time-consuming and cannot readily identify the interaction between variables[2]. Therefore, to accelerate the optimization process in an efficient, sustainable, and economical way, we created a systematically-compiled dataset of thermal polymerization to produce MP from the data mining of the literature and the results of our own experiments and employed machine learning (ML) tools to bridge the inputs and outputs, predict formation performance and properties of MP, and develop ML-based optimization for producing the desired MP. Several ML algorithms were evaluated with ten-fold cross-validation, and the algorithm with the greatest accuracy was selected. Moreover, additional detailed information behind the models was extracted. Partial dependence analysis determined the modes of each variable affecting the yield, mesophase content, and softening points of targeted MP. Further optimal solutions from reverse optimization were experimentally verified to produce MP with high mesophase content and suitable softening points. This research can provide a reference for value-added utilization of low-rank hydrocarbon resources and wastes and extend the knowledge of MP production.

References

[1] Wang WJ, Preciado I, Malzahn J, Eddings E. Mild Solvolysis Liquefaction of Low-Rank Coal into a Feedstock of Value-Added Carbon Materials. *2021 AIChE Annual Meeting.* Boston, MA: AIChE; 2021.

[2] Wang W, Preciado I, Eddings E. Thermochemical Co-Conversion of Waste Polyolefins with Low-Rank Aromatic-Rich Hydrocarbons into an Intermediate of High-Quality Anisotropic Pitch. *2021 AIChE Annual Meeting.* AIChE; 2021.

283c Bayesian Reaction Optimization of *Rac*-Lactide Polymerization Catalyzed By Aluminum Complexes

Yang Huang, Maverick Lin, Rong Tong, Hongliang Xin

Reaction optimization is experimentally costly due to the tremendous chemical search space spanned by functional materials and condition variables that determine the reaction outcomes. Recent advances in data science provide opportunities for accelerating the process via the identification of promising points in the feature space with only a small number of experiments. However, those studies largely focus on small molecule synthesis, and scarcity remains on polymerization. Herein, Bayesian optimization (BO), an iterative global optimization algorithm, is employed to fill in the gap, in a case study of *rac*-lactide polymerization catalyzed by aluminum (Al) complexes. BO model performance is tested on around 60 existing experimental data points from the literature within the overall chemical space (576 in size) built upon Al-complex catalysts with different functional groups, targeting towards the highest isotacticity (*P*m) and syndiotacticity (*P*r) values (defined from 0 to 1), two important quantities that determine polymer physical properties. The model shows a good prediction capability with mean absolute errors around 0.1. The optimization performance comparing BO with the traditional random search indicates a higher search efficiency and stability of BO. By using the model trained on these existing data points, we are currently searching through the chemical space to efficiently identify new optimal Al-complexes with high *P*m or *P*r values. Synthesizability is considered via incorporating the number of synthesis steps into the searching procedures for efficient synthesis. Two new catalysts with NMR-measured *P*m and *P*r over 0.9 were suggested by BO and later experimentally validated, indicating the effectiveness of the BO approach in finding polymerization catalysts.

283d Inferring Structure-Entropy Correlations of Zeolite-Adsorbate Interactions Using Monte-Carlo Simulations and Machine Learning

Christopher Rzepa, Srinivas Rangarajan

Zeolites are porous crystalline alumino-silicates with distinct pore topologies; and much work has been devoted to optimizing their use in catalytic applications. Advancements in ab-initio simulations have made possible to predict the catalytic properties of a zeolite framework on a particular reaction through quantification of the energy landscape, and subsequently the development of microkinetic models. Of the contributions to the free energy, entropic losses within zeolite catalysis have shown to be crucial toward rate inhibition and therefore product selectivity; and have traditionally been related with the degree of confinement associated with adsorption. Nevertheless, the adsorption entropy remains difficult to quantify, or expensive to calculate, but has been shown to follow straightforward correlations with considerable predictive capabilities.

Campbell and Sellers compiled a collection of experimental adsorbate entropies on two-dimensional catalytic surfaces.[1] The key finding of their work was that the ratio of the adsorbed-phase entropy to the gas-phase entropy was approximately two-thirds across all molecules. They proposed a simple explanation, suggesting that the adsorption of a molecule from an unhindered gas phase onto a two-dimensional surface would eliminate a dimension of translational freedom, i.e. the adsorbate behaves as a 2D gas. Dauenhauer and Abdelrhaman[2] expanded this idea to three-dimensional frameworks by compiling experimentally determined adsorption entropies for alkanes adsorbed in aluminosilicate zeolites. They showed that the entropic loss upon adsorption correlates with the molecule's gas-phase translational and rotational entropies and that the occupiable volume of a zeolite is a useful descriptor in predicting such losses.

Recently, we have implemented TraPPE forcefields[3] and quantified the adsorbate entropy by performing canonical Monte-Carlo integrations using the FEASST[4] molecular simulation package for one-hundred-eighty-five molecule-zeolite combinations.[5] Our results show that these linear correlations persist for our larger and more chemically diverse dataset; and are governed by descriptors of the zeolite's topology. The ability of our simulations to qualitatively capture the trends exhibited experimentally suggests that our simulations can be used to build predictive models based on structure-property relationships.

Herein we expand our dataset to over five-thousand molecule-zeolite combinations and implement an artificial neural network using over three-hundred framework and molecule descriptors. Our model is capable in predicting the adsorption entropies with a mean absolute error of less than the universal gas constant. Our ultimate goal is to use these highly predictable models to better understand the structure-property relationships in the adsorption entropies whilst exploiting their predictive capabilities. This talk will discuss the results of training an ANN on an expanded dataset, systematic identification of descriptors capturing the underlying correlations, as well as applying alternative, highly interpretable algebraic models.

1. Campbell, C. T.; Sellers, J. R. V. The entropies of adsorbed molecules. J. Am. Chem. Soc. 2012, 134 (43), 18109−18115.
2. J. Dauenhauer, O. A. Abdelrahman, ACS Central Science 2018 4 (9), 1235-1243
3. G. Martin and; J. Ilja Siepmann, J. Physical Chemistry B 1998 102 (14), 2569-2577
4. Hatch, H. W., Mahynski, N. A., and Shen, V. K. (2018) FEASST: Free Energy and Advanced Sampling Simulation Toolkit. J. Res. Natl Inst Stan, 123, 123004.
5. Rzepa, C., Siderius, W. D., Hatch, H. W., Shen, V. K., Rangarajan, S., and Mittal, J., J. Physical Chemistry C 2020 124 (30), 16350-16361

152a Combining functional group and Graph Neural Networks Towards Interpretable Molecular Property Models

Adem Rosenkvist Nielsen Aouichaoui, Fan Fan, Seyed Soheil Mansouri, Jens Abildskov, Gurkan Sin

Recent developments in the field of machine learning and deep learning have resulted in a surge in the integration and application of these developments in almost every field related to engineering and life science ranging from fault diagnostics in production facilities (Bao et al., 2022), to forecasting supply and demand (Hubbs et al., 2020) as well as predicting molecular properties (Wieder et al., 2020). The ability to infer the properties of molecules from their structure also known as Quantitative Structure-Property Relations (QSPRs) has long been a subject of great interest and importance in the development of computer-aided tools for process and product design as well as high throughput in-silico screening.

The introduction of the Graph Neural Network (GNN) models as an alternative modeling approach for QSPR modeling has eliminated the tedious task of developing molecular descriptors capable of expressing the molecular structural information and instead streamlined this procedure through the message passing concept (Gilmer et al., 2017). GNNs operate on molecular graphs which can be considered a natural representation of the molecular topology with nodes and edges representing the atoms and the bonds in a molecule respectively to extract a suitable representation that correlates well with the property of interest (Jiménez-Luna et al., 2020). The flexibility of these models and their ability to model properties in an end-to-end learning framework has resulted in the development of a large number of models for various applications such as predicting the critical properties (Aouichaoui et al., 2022) and various ADMET-related properties of chemicals (Yang et al., 2019).

While these models have proven they are capable of matching and exceeding the performance of descriptor-based models (Jiang et al., 2021), they come with an added disadvantage in the form of a lack of transparency and interpretability (Jiménez-Luna et al., 2020). The black-box nature of these models might hinder their wider applicability, especially in fields that rely on a first-principle understanding of the desired phenomenon. An added aspect of interpretability would potentially provide new insights into the targeted phenomenon as well as avoid the "clever-Hans effect" (provide correct predictions for the wrong reasons) (Jiménez-Luna et al., 2020).

In this work, we will present a model that combines functional groups as defined in the well-known group additivity models with GNNs. This model combines the strength of both concepts especially the interpretable aspect of group-contribution models and the flexibility and ability to extract neighborhood information of the GNN models. The constructed model is benchmarked against state-of-the-art GNN models (Gilmer et al., 2017; Xiong et al., 2020; Yang et al., 2019; Zhang et al., 2021). The results show comparable accuracy with the added aspect of interpretability, which is then shown to be consistent with chemistry and thermodynamic insights of the target properties investigated such as Aqueous solubility, enthalpy of fusion, and heat of combustion. In short, the presentation will provide the following:

- Showcase how chemistry knowledge is combined with GNN models.
- Showcase the interpretable aspect of the model
- Benchmark the performance of the model against the state-of-the-art GNN models and group-additivity models.
- Showcase the insights obtained through such models and compare them to prior knowledge

References

Aouichaoui, A.R.N., Mansouri, S.S., Abildskov, J., Sin, G., 2022. Uncertainty estimation in deep learning-based property models: Graph neural networks applied to the critical properties. AIChE Journal.

Bao, Y., Wang, B., Guo, P., Wang, J., 2022. Chemical process fault diagnosis based on a combined deep learning method. Can J Chem Eng 100, 54–66.

Gilmer, J., Schoenholz, S.S., Riley, P.F., Vinyals, O., Dahl, G.E., 2017. Neural Message Passing for Quantum Chemistry, in: 34th International Conference on Machine Learning, ICML 2017. pp. 2053–2070.

Hubbs, C.D., Li, C., Sahinidis, N. v., Grossmann, I.E., Wassick, J.M., 2020. A deep reinforcement learning approach for chemical production scheduling. Comput Chem Eng 141, 106982.

Jiang, D., Wu, Z., Hsieh, C.Y., Chen, G., Liao, B., Wang, Z., Shen, C., Cao, D., Wu, J., Hou, T., 2021. Could graph neural networks learn better molecular representation for drug discovery? A comparison study of descriptor-based and graph-based models. J Cheminform 13, 1–23.

Jiménez-Luna, J., Grisoni, F., Schneider, G., 2020. Drug discovery with explainable artificial intelligence. Nat Mach Intell.

Wieder, O., Kohlbacher, S., Kuenemann, M., Garon, A., Ducrot, P., Seidel, T., Langer, T., 2020. A compact review of molecular property prediction with graph neural networks. Drug Discov Today Technol 37, 1–12.

Xiong, Z., Wang, D., Liu, X., Zhong, F., Wan, X., Li, X., Li, Z., Luo, X., Chen, K., Jiang, H., Zheng, M., 2020. Pushing the boundaries of molecular representation for drug discovery with the graph attention mechanism. J Med Chem 63, 8749–8760.

Yang, K., Swanson, K., Jin, W., Coley, C., Eiden, P., Gao, H., Guzman-Perez, A., Hopper, T., Kelley, B., Mathea, M., Palmer, A., Settels, V., Jaakkola, T., Jensen, K., Barzilay, R., 2019. Analyzing Learned Molecular Representations for Property Prediction. J Chem Inf Model 59, 3370–3388.

Zhang, Z., Guan, J., Zhou, S., 2021. FraGAT: a fragment-oriented multi-scale graph attention model for molecular property prediction. Bioinformatics 37, 2981–2987.

152b Evaluation of Sampling Algorithms to Explore the Latent Space Created By Deep Generative Models

Fangxi Wang, Parisa Farzeen, Charles Plate, Lakshmikumar Kunche, Sanket Deshmukh

In generative chemistry, artificial intelligence based design of small molecules employs a solution to discover new chemical compounds with target properties by systematically exploring a large design space. This design space is normally explored with stochastic sampling. Thus, new candidate discovery is not only dependent on initial data but also on the sampling algorithms. In this study, different sampling methods such as metropolis-hastings, ensemble slice sampling, and multi-grid search are investigated to understand the latent space exploration performance for new molecules. Specifically, we utilize syntax-directed variational autoencoders (SD-VAE) to create a latent space for designing new drug-like molecules with strong binding energy (< -9.0 kcal/mol) to SARS-Cov-2 Spike protein. The sampling algorithms were integrated with a stacked ensemble model (SEM) for efficient exploration of the latent space. Then, the drug-like molecule candidates from different sampling algorithms are validated through docking using Autodock Vina as well by performing additional structural and pharmacophore models. The top candidates were later functionalized with functional groups based on glycomaterials to design biocompatible drug-like molecules with better binding affinities to the Spike protein. This machine learning approach may help us find good synthesizable carbohydrate-based drug-like molecules for Covid-19 treatment and might be extended to discover micromolecules for different applications.

152c Automatic Creation of Molecular Substructure Descriptors for Estimation of Pure Compound Properties

Qiong Pan, Jie Li, Xiaolei Fan

With the rapid development of chemical industries especially advances in chemical synthesis technology, the population of chemical compounds has been substantially increasing.[1] Understanding of chemical properties is an important prerequisite for effective utilization of chemical compounds.[2] In general compound properties such as normal boiling point and Reid vapour pressure are often obtained through experiments. Although it is the most reliable approach, the experimental measurement of a large number of compound properties is time-consuming and expensive.[3,4] In addition, compound properties could change when pressure or temperature changes. As a result, it is imperative to develop suitable mathematical models for prediction of compound properties using experimental data available.

The earliest method for property prediction is the use of simple correlations. For instance, a simple correlation of the property value with the number of carbon atom is developed accordingly.[5,6] However, this simple correlation method is only suitable for single family or certain common organic compounds.[7] The quantitative structure activity/property relationship (QSAR/QSPR) approach has been widely used for property estimation of chemical compounds.[2,8-10] It converts physical and chemical information of a molecule into molecular descriptors and correlates chemical properties with the numerical values of molecular descriptors. However, descriptors are generated according to various distinct rules thus interpretation of some descriptors provisional and weak.[11,12] The group contribution (GC) method which fractionates a molecule into groups in common, and then develops correlations for property estimation based on the contribution of groups to the target property. However, risk prediction could happen when the predefined groups cannot represent structure building blocks of the molecule. Then the connectivity index (CI) method[13] in which a molecule is represented by the valence connectivity indices[14] was introduced to automatically create missing groups. However, the CI method was limited due to the number of atoms available and connectivity indices.

Despite some drawbacks mentioned above, successful application of the GC and CI methods suggests molecular substructure information can be effective for property estimation. Whilst the QSPR method generates descriptors automatically, which means it could be easily adapted to the object to which it applies. Therefore, an approach that may improve the prediction should be able to extract the structural characteristics of molecules efficiently and automatically. In other words, making generation and selection of groups compelled by data analysis and interpretation, rather than by intuition or by personal experience, which meets the recently popular "data-driven" concept.

In this work, we develop a novel framework which can be divided into three parts. It includes matrix representation of molecular connectivity information, submatrix feature generation and regression model development. At first, a molecule is represented by a SMILES code, which can be directly generated using the ChemDraw software[15] if the chemical structural is known. Based on the popular cheminformatic tool RDKit, we develop a program to automatically extract atom and bond connection information from the canonical SMILES code of a molecule. With the canonical SMILES code representation of a molecule, each atom in the molecule is identified and labelled using a number sequentially. The number is also called the ID of the atom. A $N \times N$ square matrix (denoted as M=

$(a_{ij})_{(N \times N)})$ is then constructed to store the connectivity information of a molecule, where N is equal to the total number of atoms in the molecule. A diagonal entry in the matrix is used to store atomic number of an atom in the molecule with the index of the diagonal element consistent with the ID of the atom. A non-diagonal entry (i,j) is used to store bonding information between atom i and atom j. If there is not a bond connecting atom i and atom j, then the entries (i, j) and (j, i) are set to 0. In contrast, 1 for single bond, 2 for double, 1.5 for aromatic bond and 3 for triple bond. By doing this, all structural information of a molecule is stored in this matrix. Therefore, we can extract the environment of each atom in the molecule (i.e. its own information and connectivity information with other atoms). Such environment of an atom in the molecule can also be represented by a matrix, which can be obtained from the connectivity and hence a sub-matrix of M. Each environment of an atom can be considered as a feature of this atom. The environment or feature of an atom in the molecule largely depends on the length of its contiguous atoms. The length can be 1,2,3,4, or more. The corresponding chemical features generated are called the 1st-order feature, 2nd-order feature, 3rd-order feature and 4th-order feature, which are represented by different order submatrices. Lastly, feature engineering methods such as feature selection and dimensional reduction are applied, remained chemical features are used to train and test the artificial neural network (ANN) model. Pearson's feature selection and PCA reduction are implemented for data cleaning before input to the ANN for regression. ANN hyper-parameters are tuned in order to test the robustness of the proposed method.

In the case study, we adopted the largest normal boiling point database with 5276 molecules.[4] All of the molecules were input to generate molecular substructure features. Generated features of three compounds including Ortho-xylene, Meta-xylene and Para-xylene illustrate only using 1st order features, the three isomers can't be differentiated. While higher order features can be used to differentiate isomers. 15% data points were extracted as test set. Other 85% dataset were used to train the Artificial Neural Network model, K-fold cross-validation (K = 6 in this work) was implemented. The 6 trained prediction model have a close performance, which indicated the proposed method is reliable. The mean absolute percentage (MAPE) error is 4.09% on the test set. Alshehri et al.[4] developed a GPR (Gaussian Process Regression) model on GC features. As comparison, we on the one hand input GC features and the proposed submatrix features into the GPR model, the root mean squared error (RMSE) on the test set were 37.64 and 33.33 respectively, while on the training set and test set, the result RMSE were 12.07 and 9.10 respectively. On the other hand, we input both the GC features and the proposed submatrix features of test set into the developed ANN model, the MAPE and RMSE of submatrix features are 34.2177 and 4.774 respectively, which is lower than 40.441 and 5.53 of GC features. The reduced RMSE indicates the proposed submatrix feature method is accurate for property prediction of normal boiling point. The case study used in this text is estimation of normal boiling points, nevertheless, the method could be further applied to production of other chemical properties, since the method can be easily applied by only replacing the input molecules and target property.

References

1. Yoshida, J. i.; Kim, H.; Nagaki, A., Green and sustainable chemical synthesis using flow microreactors. ChemSusChem 2011, 4 (3), 331-340.
2. Katritzky, A. R.; Karelson, M.; Lobanov, V. S., QSPR as a means of predicting and understanding chemical and physical properties in terms of structure. Pure and Applied Chemistry 1997, 69 (2), 245-248.
3. Wen, H.; Su, Y.; Wang, Z.; Jin, S.; Ren, J.; Shen, W.; Eden, M., A systematic modeling methodology of deep neural network-based structure-property relationship for rapid and reliable prediction on flashpoints. AlChE J. 2021, e17402.

4. Alshehri, A. S.; Tula, A. K.; You, F.; Gani, R., Next generation pure component property estimation models: With and without machine learning techniques. AIChE J. 2021, e17469.
5. Korsten, H., Characterization of hydrocarbon systems by DBE concept. AIChE J. 1997, 43 (6), 1559-1568.
6. Fisher, C.; CH, F., Equations correlate n-alkane physical properties with chain length. 1982.
7. Van Nes, K.; Van Westen, H. A., Aspects of the constitution of mineral oils. Elsevier Publishing Company: 1951.
8. Vom Lehn, F.; Brosius, B.; Broda, R.; Cai, L.; Pitsch, H., Using machine learning with target-specific feature sets for structure-property relationship modeling of octane numbers and octane sensitivity. Fuel 2020, 281, 118772.
9. Shi, X.; Li, H.; Song, Z.; Zhang, X.; Liu, G., Quantitative composition-property relationship of aviation hydrocarbon fuel based on comprehensive two-dimensional gas chromatography with mass spectrometry and flame ionization detector. Fuel 2017, 200, 395-406.
10. Roubehie Fissa, M.; Lahiouel, Y.; Khaouane, L.; Hanini, S., QSPR estimation models of normal boiling point and relative liquid density of pure hydrocarbons using MLR and MLP-ANN methods. J. Mol. Graphics Modell. 2019, 87, 109-120.
11. Todeschini, R.; Consonni, V., Handbook of molecular descriptors. Wiley-VCH, Weinheim: 2000.
12. Dong, J.; Yao, Z. J.; Zhang, L.; Luo, F.; Lin, Q.; Lu, A. P.; Chen, A. F.; Cao, D. S., PyBioMed: a python library for various molecular representations of chemicals, proteins and DNAs and their interactions. J Cheminform 2018, 10 (1), 16.
13. Gani, R.; Harper, P. M.; Hostrup, M., Automatic creation of missing groups through connectivity index for pure-component property prediction. Industrial andamp; Engineering Chemistry Research 2005, 44 (18), 7262-7269.
14. Kier, L. B.; Hall, L. H., Molecular connectivity in structure-activity analysis. Research studies: 1986.
15. Cousins, K. R., Computer review of ChemDraw ultra 12.0. ACS Publications: 2011.

152d *Data-Driven Chemical Property Models for Energetic Materials Using Transfer Learning*

Joshua Lansford, Brian C. Barnes, Betsy Rice, Klavs Jensen

Due to the hazardous nature of energetic materials, it is useful to have accurate estimates of physical properties related to their handling, such as impact sensitivity and vapor pressure. Unfortunately, many safety-related properties depend on multiscale interactions and cannot be directly computed with high accuracy. By themselves, physics-based property prediction models do not extrapolate well and can fail entirely. While machine learning (ML) can overcome these limitations, ML requires large datasets that are not available for energetic properties. Here, we apply two different transfer learning approaches to predict impact sensitivity and vapor pressure. In the first approach, model parameters are learned to map a chemical graph to properties that can be directly computed, and then these parameters are used to predict impact sensitivity. Specifically, we co-train a directed-message passing neural network on a diverse dataset in order to predict impact sensitivity. In the second approach, we embed a physical model into the neural network to enable extrapolation and improve out-of-sample prediction accuracy for energetic vapor pressures. Our models outperform existing models on a diverse test set and are generalizable.

152e Machine-Learning a Solution for Reactive Simulations of Complex Chemical Systems

Rebecca Lindsey, Laurence E. Fried, Nir Goldman, C. Huy Pham, Sorin Bastea

Many grand challenges in computational chemistry involve accurate description of condensed-phase reactivity. For example, understanding chemistry at extreme conditions (e.g. 1000s of K and 10s of GPa) is crucial in fields spanning geochemistry, astrobiology, and alternative energy. Quantum mechanical methods like Kohn-Sham density functional theory (DFT) can provide valuable microscopic insights into such systems while circumventing the risks of physical experiments, but the associated reactive scales (i.e. ns and μm) largely preclude extension of such models to molecular dynamics. These limitations have motivated extensive work towards development of scalable classical reactive interatomic potentials; however, the complicated forms underlying these models require time consuming parameterization, yet generally still fail to capture the physics of chemistry under extreme conditions. As a result, the physiochemical space over which accurate reactive interatomic potential parameters exist is limited.

In this work, we present the Chebyshev Interaction Model for Efficient Simulation (ChIMES), a machine-learned generalized many-body reactive interatomic potential. ChIMES models are constructed from linear combinations of Chebyshev polynomials, and are rapidly generated by force, stress, and energy matching to short DFT trajectories. These models retain most of the accuracy of DFT while decreasing computational requirements by several orders of magnitude and enabling linear-scaling with respect to system size. To date, these models have been applied to condensed-phase reaction-driven problems including shockwave-driven nano-carbon formation, hydride embrittlement, and equation-of-state prediction. Details of the model and fitting approach will be discussed along with selected applications.

This work was performed under the auspices of the U.S. Department of Energy by Lawrence Livermore National Laboratory under Contract DE-AC52-07NA27344. LLNL-ABS- 808405

152f Capturing Molecular Interactions in Graph Neural Networks: A Case Study in Multi-Component Phase Equilibrium

Shiyi Qin, Shengli Jiang, Prasanna Balaprakash, Reid Van Lehn, Victor Zavala

Machine learning has been widely to predict diverse molecular properties such as water solubility, toxicity, and lipophilicity [1]. In these approaches, molecular descriptors or fingerprints are obtained as the input data to develop quantitative structure-property relationship (QSPR) models [2,3]. More recently, there has been an increasing trend in applying deep learning architectures to study more complex chemical systems with multiple components, such as alloys [4], copolymers [5,6], chemical reactions [7,8], and gas [9] and liquid [10,11,12] mixtures. Among these deep-learning techniques, graph neural networks (GNNs) [13] have gained special popularity because they can directly use molecular graph representations, thus avoiding the need to pre-calculate/pre-define molecular descriptors.

In a general GNN-based approach for molecular property prediction, atom and bond features are propagated based on the molecular structure for a single molecule input. The embedded features are then sent to fully-connected layers to construct predictive models [14]. When dealing with multiple components, several attempts have been made. The typical method is to average or concatenate the embedded features of individual molecules and use them as the system-level features for property inference with fully-connected or attentive layers [6,7,8]. Previous studies have also incorporated weighted sums or concatenation to take into account the composition information when needed [6]. However, these approaches have not captured intra- and inter- molecular interactions in an explicit manner.

In this work, we present a GNN architecture to incorporate both intra- and inter- molecular interactions via the combination of atomic-level (local) graph convolution and molecular-level (global) message passing for property prediction of multi-component chemical systems. To connect local features with global features, we constructed a molecular interaction network as the intermediate step. The molecular interaction network is a complete graph with each composition-weighted node representing a molecule and each edge representing a hypothetical inter-molecular interaction, such as hydrogen bonding information. It serves as a physics-informed topological prior to aid feature extraction from multi-component systems. Here, we tested the proposed GNN architecture through a case study on activity coefficient predictions of multi-component systems. We also provided a framework that can intake a given mixture (binary or ternary) and generate the corresponding phase diagrams (P-x-y) using the trained GNN along with thermodynamic calculations. We also performed counter-factual analysis [15] of the trained model to identify the impact of functional groups on activity coefficients to obtain physical insights.

References

[1] Sanchez-Lengeling, B., andamp; Aspuru-Guzik, A. (2018). Inverse molecular design using machine learning: Generative models for matter engineering. Science, 361(6400), 360-365.

[2] Rogers, D., andamp; Hahn, M. (2010). Extended-connectivity fingerprints. Journal of chemical information and modeling, 50(5), 742-754.

[3] Karelson, M., Lobanov, V. S., andamp; Katritzky, A. R. (1996). Quantum-chemical descriptors in QSAR/QSPR studies. Chemical reviews, 96(3), 1027-1044.

[4] Natarajan, A. R., andamp; Van der Ven, A. (2018). Machine-learning the configurational energy of multicomponent crystalline solids. npj Computational Materials, 4(1), 1-7.

[5] Wilbraham, L., Sprick, R. S., Jelfs, K. E., andamp; Zwijnenburg, M. A. (2019). Mapping binary copolymer property space with neural networks. Chemical science, 10(19), 4973-4984.

[6] Hanaoka, K. (2020). Deep neural networks for multicomponent molecular systems. ACS omega, 5(33), 21042-21053.

[7] Wei, J. N., Duvenaud, D., andamp; Aspuru-Guzik, A. (2016). Neural networks for the prediction of organic chemistry reactions. ACS central science, 2(10), 725-732.

[8] Coley, C. W., Jin, W., Rogers, L., Jamison, T. F., Jaakkola, T. S., Green, W. H., ... andamp; Jensen, K. F. (2019). A graph-convolutional neural network model for the prediction of chemical reactivity. Chemical science, 10(2), 370-377.

[9] Pan, Y., Ji, X., Ding, L., andamp; Jiang, J. (2019). Prediction of lower flammability limits for binary hydrocarbon gases by quantitative structure—property relationship approach. Molecules, 24(4), 748.

[10] Wang, T., Tang, L., Luan, F., andamp; Cordeiro, M. N. D. (2018). Prediction of the toxicity of binary mixtures by QSAR approach using the hypothetical descriptors. International journal of molecular sciences, 19(11), 3423.

[11] Chinta, S., andamp; Rengaswamy, R. (2019). Machine learning derived quantitative structure property relationship (QSPR) to predict drug solubility in binary solvent systems. Industrial andamp; Engineering Chemistry Research, 58(8), 3082-3092.

[12] Jirasek, F., Alves, R. A., Damay, J., Vandermeulen, R. A., Bamler, R., Bortz, M., ... andamp; Hasse, H. (2020). Machine learning in thermodynamics: Prediction of activity coefficients by matrix completion. The Journal of Physical Chemistry Letters, 11(3), 981-985.

[13] Wu, Z., Ramsundar, B., Feinberg, E. N., Gomes, J., Geniesse, C., Pappu, A. S., ... andamp; Pande, V. (2018). MoleculeNet: a benchmark for molecular machine learning. Chemical science, 9(2), 513-530.

[14] Zhou, J., Cui, G., Hu, S., Zhang, Z., Yang, C., Liu, Z., ... andamp; Sun, M. (2020). Graph neural networks: A review of methods and applications. AI Open, 1, 57-81.

[15] Wellawatte, G. P., Seshadri, A., andamp; White, A. D. (2022). Model agnostic generation of counterfactual explanations for molecules. Chemical Science.

152g A Statistical Analysis of Computational Hydrogen Bond Features in Hydrophobic Non-Ion Deep Eutectic Solvents and Non-Deep Eutectic Solvents

Usman Abbas, Joseph Tapia, Yuxuan Zhang, Mohammad Selim, Jian Shi, Jin Chen, Qing Shao

Hydrophobic non-ion deep eutectic solvents have shown promise as advanced extractants for separate pollutants from aqueous solutions because of their potential in low toxicity, extremely low water solubility, and economic availability. However, the current discovery of hydrophobic non-ion deep eutectic solvents mainly relies on the intuition of the researchers or a trial-and-error method, which may have a low success rate and neglect many promising candidates. Hydrogen bonds between the compounds are considered the primary reason for the deep eutectic effect. Thus, the hydrogen bonds are likely to be the source of proper features that can be used to design new deep eutectic solvents and predict their properties. To guide the development of hydrogen bond-based deep eutectic solvent design, we analyze the structural and dynamic properties of hydrogen bonds in 50 hydrophobic deep eutectic solvents and 100 non deep eutectic solvent systems formed by various organic hydrogen bond donors and acceptors using molecular dynamics simulations. The structural and dynamic properties of the hydrogen bonds are analyzed using the distributions of average numbers and lifetimes of the inter- and intra- hydrogen bonds between compounds. The statistical analysis shows that the hydrophobic deep eutectic solvents possess a distinct gap between the compounds' average intra-hydrogen bond numbers that the non-deep eutectic solvent systems do not present. This statistical analysis indicates that the intra-component and inter-component hydrogen bond numbers could be used to develop features that differentiate deep eutectic solvents from the non-deep eutectic solvent systems. Based on this observation, we developed a machine learning model to classify if two organic molecules can form deep eutectic solvents.

152h Predicting Temperature-Dependent Activity Coefficients of Ionic Liquid-Solute Systems through Graph-Based Machine Learning

Jan G. Rittig, Karim Ben Hicham, Artur M. Schweidtmann, Manuel Dahmen, Alexander Mitsos

Ionic liquids (IL) have attractive physicochemical properties and are highly interesting solvents in the design of sustainable separation processes [1,2]. Properties of ILs can be adapted to the separation task at hand by combining anions, cations, and further structural groups so that the number of potential IL candidates is large including at least one million possible ILs [2]. To assess the suitability of an IL as a solvent candidate, the activity coefficient is of paramount importance. Compared to the large number of potential ILs, the experimental data on activity coefficients for ILs is, however, limited which calls for prediction models. Classical activity coefficient prediction models such as UNIFAC and COSMO-RS are well established in chemical engineering but are limited in terms of applicability and accuracy for IL-solute systems [3].

Recently, machine learning models have shown promising results for activity coefficient prediction, with two methods being actively investigated: Matrix completion methods (MCMs) and graph neural networks (GNNs) [3-6]. MCMs build on the idea that the task of activity coefficient prediction can be represented as filling entries of a matrix with rows as solutes and columns as solvents (or vice versa) and with each entry corresponding to the activity coefficient of a pair of solute and solvent [3,4]. MCMs have been applied to predict activity coefficients of IL-solute systems, showing superior accuracy compared to classical thermodynamic models specifically adapted to ILs like UNIFAC-IL and calibrated COSMO-RS [3]. A GNN, on the other hand, is a graph-based machine learning method that has shown great results in predicting physicochemical properties of molecules in recent years [5,6]. GNNs operate on the molecular graph with atoms as nodes and bonds as edges and thereby enable learning a direct mapping from the molecular graph to a molecular property of interest. Very recently, GNNs have been utilized for predicting activity coefficients of solvent-solute systems at a constant reference temperature with promising results [7,8].

We propose a GNN for predicting the temperature-dependent activity coefficient at infinite dilution of IL-solute systems. The proposed model extends GNNs for activity coefficient prediction to ILs and temperature-dependent activity coefficient values. To compare our GNN to state-of-the-art MCM models for IL-solute systems, we also implement an MCM approach and train both models on a large database including more than 40,000 data points [3,9]. The results show that GNNs and MCMs achieve comparable high prediction quality for predicting the activity coefficient of IL-solute systems. Whereas the applicability of MCMs is limited to ILs and solute molecules that have been included in model training, GNNs can be applied to ILs and solutes not seen during training. Our investigations show that the GNN allows for generalization with high accuracy, making it a promising tool for computer-aided design of ILs for chemical engineering applications. An open-source publication of the models and code is underway.

[1] Han, X., andamp; Armstrong, D. W. (2007). Ionic liquids in separations. Accounts of Chemical Research, 40(11), 1079-1086.

[2] Rogers, R. D., andamp; Seddon, K. R. (2003). Ionic liquids – Solvents of the future?. *Science*, *302*(5646), 792-793.

[3] Chen, G., Song, Z., Qi, Z., andamp; Sundmacher, K. (2021). Neural recommender system for the activity coefficient prediction and UNIFAC model extension of ionic liquid-solute systems. AIChE

Journal, 67(4), e17171.

[4] Jirasek, F., Alves, R. A., Damay, J., Vandermeulen, R. A., Bamler, R., Bortz, M., Mandt, S., Kloft, M. andamp; Hasse, H. (2020). Machine learning in thermodynamics: Prediction of activity coefficients by matrix completion. The Journal of Physical Chemistry Letters, 11(3), 981-985.

[5] Medina, E. I. S., Linke, S., Stoll, M., andamp; Sundmacher, K. (2022). Graph neural networks for the prediction of infinite dilution activity coefficients. Digital Discovery. Advance online publication.

[6] Felton, K. C., Ben-Safar, H., Alexei, A. A. (2022). DeepGamma: A deep learning model for activity coefficient prediction. 1st Annual AAAI Workshop on AI to Accelerate Science and Engineering (AI2ASE), 02-28-2022, virtual.

[7] Gilmer, J., Schoenholz, S. S., Riley, P. F., Vinyals, O., andamp; Dahl, G. E. (2017). Neural message passing for quantum chemistry. Proceedings of the 34th International Conference on Machine Learning, in Proceedings of Machine Learning Research 70:1263-1272.

[8] Wieder, O., Kohlbacher, S., Kuenemann, M., Garon, A., Ducrot, P., Seidel, T., andamp; Langer, T. (2020). A compact review of molecular property prediction with graph neural networks. Drug Discovery Today: Technologies, 37, 1-12.

[9] Kazakov, A., Magee, J.W., Chirico, R.D., Paulechka, E., Diky, V.; Muzny, C.D., Kroenlein, K., Frenkel, M. (2013). NIST Standard Reference Database 147: NIST Ionic Liquids Database - ILThermo v2.0, National Institute of Standards and Technology, Gaithersburg, USA, https://ilthermo.boulder.nist.gov (accessed on 4-4-2022).

152i Property Prediction of Amine-Functionalized Ionic Liquids for Multi-Scale Carbon Capture Design and Optimization

Austin Keller, Pratik Kelkar, Michael Baldea, Mark Stadtherr, Joan Brennecke

Atmospheric carbon dioxide remediation is one of the great challenges of the current generation. However, carbon-intensive processes are likely to remain industrially relevant. For this reason, efficient point source carbon capture technologies are a vital tool to slow the increase in atmospheric CO_2 concentration. Aqueous amine-based solvents have been proposed as one such technology,[1] but suffer from toxic emissions due to solvent volatility and degradation,[2] as well as heat intensive regeneration due to unfavorable reaction thermodynamics[3] and solvent and water evaporation[4]. Aprotic *N*-heterocyclic anion-based ionic liquids (AHA ILs) are a class of ionic liquids that reversibly react with CO_2 and that have been proposed as a substitute for aqueous amines due to their high thermal stability, low corrosivity, negligible vapor pressure, and tunability.

In particular, the large variety of AHA ILs that may be synthesized presents an opportunity to find optimal candidates for each of the many CO_2 capture applications, covering a wide range of CO_2 concentrations. This presents an ideal opportunity for combined materials/process optimization, however, the expansive design space and nonlinear relationships for these compounds are difficult for the designer to navigate. Many of the most influential IL properties for the performance of a CO_2 capture operation—namely CO_2 capacity, density, viscosity, and heat capacity[5]—can be predicted reasonably well with machine learning models for ILs in general, but these models tend to perform inadequately when applied to the AHA IL dataset.

In this work, we investigate the relationship between the relevant CO_2 capture properties of AHA ILs with tetra-alkylphosphonium cations using quantum chemical and group contribution-based descriptors and supervised learning methods. We have data-mined and curated the relevant thermophysical property and CO_2 uptake data as well as performed and experimentally validated quantum chemical calculations to augment the CO_2 uptake dataset. Finally, we probe the effect of various dimension reduction techniques and machine learning methods on model performance and identify an optimal set of descriptors and a model for each property of interest. This will facilitate a future study in combined process and materials optimization to simultaneously find the optimal IL absorbent and capture process design for a given application.

(1) Leung, D. Y. C.; Caramanna, G.; Maroto-Valer, M. M. An Overview of Current Status of Carbon Dioxide Capture and Storage Technologies. *Renewable and Sustainable Energy Reviews* **2014**, *39*, 426–443. https://doi.org/10.1016/j.rser.2014.07.093.

(2) Koornneef, J.; Ramirez, A.; van Harmelen, T.; van Horssen, A.; Turkenburg, W.; Faaij, A. The Impact of CO2 Capture in the Power and Heat Sector on the Emission of SO2, NOx, Particulate Matter, Volatile Organic Compounds and NH3 in the European Union. *Atmospheric Environment* **2010**, *44* (11), 1369–1385. https://doi.org/10.1016/j.atmosenv.2010.01.022.

(3) Shiflett, M. B.; Drew, D. W.; Cantini, R. A.; Yokozeki, A. Carbon Dioxide Capture Using Ionic Liquid 1-Butyl-3-Methylimidazolium Acetate. *Energy Fuels* **2010**, *24* (10), 5781–5789. https://doi.org/10.1021/ef100868a.

(4) Hong, B.; Simoni, L. D.; Bennett, J. E.; Brennecke, J. F.; Stadtherr, M. A. Simultaneous Process and Material Design for Aprotic N-Heterocyclic Anion Ionic Liquids in Postcombustion CO_2 Capture. *Ind. Eng. Chem. Res.* **2016**, *55* (30), 8432–8449. https://doi.org/10.1021/acs.iecr.6b01919.

(5) Seo, K.; Tsay, C.; Hong, B.; Edgar, T. F.; Stadtherr, M. A.; Baldea, M. Rate-Based Process Optimization and Sensitivity Analysis for Ionic-Liquid-Based Post-Combustion Carbon Capture. *ACS Sustainable Chem. Eng.* **2020**, *8* (27), 10242–10258. https://doi.org/10.1021/acssuschemeng.0c03061.

152j Machine-Learning Enabled Screening of MOFs for Ion Selective Membranes

Shuwen Yue, Aditya Nandy, Heather Kulik

The design of ion selective materials with improved separations efficacy and efficiency is critical for the development of novel membrane technologies in water purification and desalination. Metal-organic frameworks (MOFs) serve as promising materials for ion separation membranes given their rigid structure and high tunability, allowing for tailored design of pore sizes, geometries, and chemical topology. Inspired by the ultra-high selectivity exhibited in biological ion channels, we aim to understand the fundamental design rules in material structure and chemical patterning which leads to these biomimetic functionalities. Towards these goals, we screen the Computationally-Ready Experimental (CoRE) MOF database of 9,598 experimentally known and synthesizable MOFs to find a set of promising materials for membranes based on structural and chemical characteristics, water stability, and ion transport. In order to determine water stability of MOF candidates, we undertake a transfer learning approach to expand existing machine-learning MOF thermal stability and solvent removal stability models to water stability. Ion-MOF binding energies calculated from DFT and ion transport properties obtained from classical molecular dynamics simulations provide further insight towards the underlying driving forces for ion selectivity and transport. Ultimately, a curated set of promising MOF candidates for ion selective membranes is identified for deeper experimental and computational investigation.

88a A Bayesian Experimental Design Framework for Optimizing Microbial Communities

Jaron Thompson, Ophelia Venturelli, Victor Zavala

Microbial communities have enormous functional potential, including the ability to valorize biofuel production [1,2], improve agricultural yields [3], and detoxify waste [4]. However, designing microbial communities to perform desired functions remains a technical challenge. While communities of microbial species can outperform individual species at industrially relevant tasks, determining such communities and optimizing operating conditions is difficult due to poorly understood mechanisms and the inability to experimentally observe all possible conditions. Recently, data-driven approaches to rationally select for microbial communities have emerged as a promising avenue for microbiome engineering [5]. Despite a handful of initial successes in data driven approaches to optimize microbial communities [6,7], these studies have relied on fitting simple ecological models to small, noisy data and did not leverage model uncertainty to maximize the information content of experimental designs. Improving the ability to engineer microbial communities will therefore require advancements in model development, parameter estimation, and optimal experimental design.

In summary, in this work, we present a Bayesian design-of-experiments framework for modeling and optimizing microbial communities directly from data. Our framework includes a recurrent neural network (RNN) architecture tailored to model microbial community dynamics directly from data, a Bayesian inference method for parameter estimation and quantification of prediction uncertainty, and a model-guided optimization approach to select batches of microbial communities that collectively maximize information content and functional design objectives.

Interactions between microbial species are complex and currently not well understood, which necessitates flexible modeling approaches that learn how species interact from experimental data. Machine learning methods that model the dynamics of microbial species abundance over time, such as recurrent neural networks, are thus compelling approaches; however, they can produce physically unrealistic predictions, such as negative species abundances or the prediction of a species appearing despite not being initially present in the community. To overcome this limitation, we present an RNN that we call Microbiome Recurrent Neural Network (MRNN), a modified RNN architecture that eliminates the possibility of predicting physically unrealistic species abundances and metabolite concentrations. Because biological data sets are typically limited to a small collection of noisy observations, we present a rigorous, automated approach to optimize the degree of model regularization to avoid over-fitting [8]. Furthermore, data acquisition is often time-consuming and expensive. Consequently, the selection of an informative set of experiments is crucial for developing models that capture system properties while minimizing time and resources spent on performing experiments [9]. We adopt a Bayesian experimental design strategy to optimize dynamic biological systems using RNNs.

Once fit to experimental data, the MRNN is used to make probabilistic predictions of previously unobserved experimental conditions. Using a utility function that balances exploration with exploitation, a subset of experimental conditions can be proposed for future experiments. We show that the MRNN outperforms a more flexible machine learning model in the prediction of species abundances over time using ground truth species abundance data simulated from a generalized Lotka-Volterra model. We then show that the MRNN accurately predicts confidence intervals of species abundance and metabolite concentration for an experimental dataset that contains 95 different microbial communities composed of unique subsets of 25 species. To demonstrate the ability of the framework to seek informative experimental designs that optimize a time-dependent objective, we show that the MRNN can be applied to increase the abundance of a set of beneficial microbial species using a ground truth resource-competition model.

References:

[1] Scarborough MJ, Lynch G, Dickson M, McGee M, Donohue TJ, Noguera DR. Increasing the economic value of lignocellulosic stillage through medium-chain fatty acid production. Biotechnology for biofuels. 2018;11(1):1–17.

[2] Agler MT, Spirito CM, Usack JG, Werner JJ, Angenent LT. Chain elongation with reactor microbiomes: upgrading dilute ethanol to medium-chain carboxylates. Energy andamp; Environmental Science. 2012;5(8):8189–8192.

[3] Kaul S, Choudhary M, Gupta S, Dhar MK. Engineering host microbiome for crop improvement and sustainable agriculture. Frontiers in Microbiology. 2021;12:1125.

[4] Loffler FE, Edwards EA. Harnessing microbial activities for environmental cleanup. Current Opinion in Biotechnology. 2006;17(3):274–284.

[5] Lawson CE. Retooling Microbiome Engineering for a Sustainable Future. Msystems. 2021;6(4):e00925–21.

[6] Clark RL, Connors BM, Stevenson DM, Hromada SE, Hamilton JJ, Amador-Noguez D, et al. Design of synthetic human gut microbiome assembly and butyrate production. Nature communications. 2021;12(1):1–16.

[7] Stein RR, Tanoue T, Szabady RL, Bhattarai SK, Olle B, Norman JM, et al. Computer-guided design of optimal microbial consortia for immune system modulation. Elife. 2018;7:e30916.

[8] Bishop CM. Pattern recognition. Machine learning. 2006;128(9).

[9] Box GE, Lucas H. Design of experiments in non-linear situations. Biometrika. 1959;46(1/2):77–9

88b Solution Coating of Polymer Blend Libraries for High-Throughput Experimentation Via Passive Mixing

Aaron L. Liu, Ezgi Melis Dogan-Guner, Michael McBride, Rahul Venkatesh, Miguel A. Gonzalez, Jun Amano, Elsa Reichmanis, Martha Grover, J Carson Meredith

As many research methodologies traditionally demand a considerable investment in time and resources, materials informatics approaches have emerged as a key development focus in enabling accelerated materials discovery. Such approaches, which often leverage machine learning and advanced data analytics techniques, require sufficiently large experimental datasets. In some materials systems, high-throughput experimentation methods can potentially bridge this small data gap by accelerating the curation of laboratory data. For high-throughput experimentation of polymer materials, one protocol is to enable the deposition of thin-film libraries that can be rapidly characterized within the parameter space of interest. However, the need to overcome unfavorable solubility constraints at ambient conditions, high viscosities, and other processing challenges can impede the ability to generate such polymer libraries. To overcome these obstacles, this work demonstrates an automated flow system capable of operation under aggressive solvent conditions up to 120°C for the solution casting of polymer thin-films. The system includes a custom-designed passive mixer suitable for effective mixing of polymer solutions under solvent use or elevated temperatures. Residence time distribution modeling was leveraged to successfully inform flow conditions and generate composition gradient films based on poly(3-hexylthiophene), poly(styrene), and poly(propylene). These composition libraries were interrogated using high-throughput property measurements and mapped to spatially resolved composition measurements. Results display the feasibility of our gradient film approach as a key component within informatics-driven workflows. Integration of data science techniques on such curated data can facilitate the discovery and development of high-performance polymer coatings, plastics, and clean energy materials, all of which often display the processing challenges that are traditionally difficult for high throughput experimentation.

88d Design of an Automatic Platform for Machine-Learning Model Based Molecular Property Optimization

Matthew McDonald, Brent Koscher, Richard Canty, Seung Kyun Ha, Camille Bilodeau, Kevin P. Greenman, Charles J. McGill, Rafael Gomez-Bombarelli, William Green, Klavs Jensen

Developments in machine learning have led to models capable of predicting a range of interesting molecular properties. Similarly, predictive retrosynthesis tools trained on large reaction databases have enabled automated planning of routes to novel compounds. Combining property and synthesis prediction, along with molecule generative modeling, allows automation of the molecule discovery process *in silico*. Physical evaluation of property-optimized molecules is a widely known bottleneck in the molecular discovery space. We have attempted to overcome this bottleneck by building a robotic platform guided by machine learning models. The platform uses a machine learning based retrosynthetic planner to synthesize interesting molecules (those having top-performing or uncertain predicted properties) and to measure and give feedback on their properties, for optimization in an automated, iterative fashion. We target three properties—optical absorbance, water/octanol partitioning, and oxidative stability—and use a scaffold approach to discover new compounds in a well-plate-based molecular design-make-test cycle.

As target molecules deviate more from the training data, the predictive power of the machine learning models decreases. To combat this, we have divided the discovery workflow into two phases: an exploration phase synthesizes a variety of molecules based on each scaffold to learn the particulars of said family, then, after retraining the models with the results of the first phase, an exploitation iteration creates a selection of high performers. Along the way, troublesome reactions are automatically optimized, with the results of optimization similarly being used to improve future synthetic routes. This talk will discuss the entirety of the workflow with respect to an example scaffold whose optical and partitioning properties have been optimized without sacrificing photo-oxidative stability. Automation efforts will be emphasized—particularly reaction execution and optimization in well plates, purification and isolation, product characterization, and engineering an interface between machine learning models and robotic hardware.

88e Simulating Closed-Loop Catalyst Discovery Processes Using an Experimental Band Gap Surrogate Model

Kirby Broderick, Zachary Ulissi

Heterogeneous catalyst discovery faces many challenges, some of which are common to materials discovery efforts: a curse of dimensionality arises from combinatorial design spaces; a lack of open-source datasets and plethora of experimental techniques make it difficult to predict experimental synthesizability or activity; and high-throughput synthesis techniques outpace structural characterization, meaning that the structures or properties models rely on often need to be inferred. Other factors are unique to catalyst discovery: because heterogeneous catalysis is a surface phenomenon, activity and stability must be considered at the level of the surface, and modifications made to increase surface area or stability often have impacts on activity that are hard to model computationally. Despite these difficulties, computational tools have proven useful in predicting catalyst synthesizability and activity in various settings, predicting effective catalysts as obscure as MoS_2 and Pt_3Sc on the basis of purely computational descriptors. However, closed-loop optimization processes that develop models of catalyst activity and iteratively acquire new experiments based on their expected utility have proven challenging to develop.

To better understand closed-loop catalyst discovery processes, I developed an optimization process to acquire experimental band gaps from a database on the Materials Project website. Crystal structures corresponding to experimental compositions were taken from the Materials Project database, and prototype structures were created in cases where there were no structures, and a graph model was used to predict various materials properties from these structures. An acquisition model using these structural features as well as composition features was then used to iteratively discover the experimental data set. Methods and knowledge developed by this effort are then applied to a simulated catalyst discovery problem with a similar structure.

88f Improved Characterization of Membrane Transport Properties through Advanced Data Analytics

Xinhong Liu, Jonathan Ouimet, Laurianne Lair, William Phillip, Alexander Dowling

Membrane processes have shown promise for addressing the critical needs for sustainability and energy efficiency. Recent material design to achieve separations of similar-sized molecules has evolved in the directions of precisely controlling the nanostructure of membranes and identifying chemical functionalities which accentuate desired transport properties. [1,2] A detailed understanding of the underlying thermodynamic and transport phenomena can elucidate the molecular interactions and mechanisms that affect the macroscopic transport properties of the membrane. [3,4] Motivated by this need, the development of membrane characterization techniques that explore the dependency of membrane performance on feed conditions can greatly accelerate the development of materials. [5] In addition, membrane characterization that elucidates underlying mechanisms provides essential information for scale-up, design, and optimization, facilitating the development of new separation systems.

Design of Experiments (DoE) methods enable optimization of computational and physical experiments that maximize the information gain and minimize time and resource costs. Classical 'black-box' DoE approaches (a.k.a. factorial, response surface), which decide the best design by the input-output relationship, do not (directly) incorporate membrane science knowledge; in contrast, model-based DoE (MBDoE) leverages high-fidelity models constructed from underlying physical principles that describe the experimental system. [6] The information collected from experiments can be applied to discriminate between scientific hypotheses, posed as mathematical models, and to improve the precision of parameter estimation. The emergence of techniques within MBDoE has great potential in the design of instruments and experimental conditions to better characterize the performance of separation devices as a function of solute concentration and in complex feed streams. However, to date, their application to problems in membrane science remains limited.

We recently proposed the Diafiltration Apparatus for high-Throughput Analysis (DATA), which enables a 10-times reduction in the time, realized with fewer experiments necessary to characterize membrane transport properties. [7] In follow-up work, we mathematically quantified these improvements in the form of information gain and further refine the static experimental conditions needed in DATA to characterize membrane transport properties. [8] In this talk, we apply the Fisher information matrix (FIM) analyses and MBDOE to further improve DATA. We highlight two non-ideal phenomena, namely "lag" and "overflow", which occur when changing the operating pressure of the system. Guided by the tools of data science, we show that modeling these phenomena can leverage the additional data within the start-up process to elucidate the underlying physics, improve the parameter precision, and brings insights to design a time-varying applied pressure in DATA. A time correction for permeate product collected is also introduced to improve the model predictions. Moreover, our framework, which integrates data analytics and instrumentation design can be applied to investigate concentration-dependent membrane performance to further accelerate the development of materials. For example, we apply the improved DATA to explore the dependency of membrane transport parameters on feed conditions of a surface-charged membrane by ranking candidate models using information criteria.

References

1. Hoffman, J. R., andamp; Phillip, W. A. (2020). 100th anniversary of macromolecular science

viewpoint: integrated membrane systems. *ACS Macro Letters*, *9*(9), 1267-1279.
2. Sadeghi, I., Kaner, P., andamp; Asatekin, A. (2018). Controlling and expanding the selectivity of filtration membranes. *Chemistry of Materials*, *30*(21), 7328-7354.
3. Geise, G. M., Paul, D. R., andamp; Freeman, B. D. (2014). Fundamental water and salt transport properties of polymeric materials. *Progress in Polymer Science*, *39*(1), 1-42.
4. Yaroshchuk, A., Bruening, M. L., andamp; Zholkovskiy, E. (2019). Modelling nanofiltration of electrolyte solutions. *Advances in Colloid and Interface Science*, *268*, 39-63.
5. Ghosh, R., andamp; Cui, Z. (2000). Analysis of protein transport and polarization through membranes using pulsed sample injection technique. *Journal of Membrane Science*, *175*(1), 75-84.
6. Franceschini, G., andamp; Macchietto, S. (2008). Model-based design of experiments for parameter precision: State of the art. *Chemical Engineering Science*, *63*(19), 4846-4872.
7. Ouimet, J. A., Liu, X., Brown, D. J., Eugene, E. A., Popps, T., Muetzel, Z. W., ... andamp; Phillip, W. A. (2022). DATA: Diafiltration Apparatus for high-Throughput Analysis. *Journal of Membrane Science*, *641*, 119743.
8. Liu, X., Wang, J., Ouimet, J. A., Phillip, W., Dowling, A. (2022) Membrane characterization with model-based design of experiments. In *14th International Symposium on Process Systems Engineering*.

88g Autonomous Optimization of Robotic Process Parameters for Accurate Experiments

Martin Seifrid

In the form of self-driving labs, automation and machine learning provide powerful tools to accelerate the pace at which we can make discoveries, and optimize processes and materials. By minimizing human involvement in performing routine tasks, self-driving labs can liberate researchers to generate hypotheses and answer important scientific questions.

Designing a self-driving lab requires balancing generality and accuracy. For example, self-driving labs designed to dose a wide range of sample quantities struggle to do so accurately for very small amounts. As a result, this limits the lab's ability to carry out experiments that require accurate dosing of small quantities of , for example, catalysts. It would therefore be highly desirable to be able to extend the operating range of robotic systems by finding optimal process parameters, potentially even exceeding manufacturer specifications. However, tuning the process parameters to find the optimal conditions is quite tedious, and significantly encroaches on valuable researcher time. By establishing fully autonomous experiments for process parameter optimization, we can remove the necessity for human involvement and maximize research efficiency.

I will discuss how a closed-loop system integrating a multi-objective Bayesian optimization algorithm and a robotic synthesis platform can be trained to dose small quantities with good accuracy and precision. Building a self-driving lab requires delicately balancing generality, specificity and equipment cost. For instance, the manufacturer specified accuracy of an instrument may limit exploration of future process of experiment parameters, while a robot capable of accuracy in a wide operating range is typically very expensive. In conjunction with a machine learning algorithm, the robot can learn to optimize its own process parameters to achieve accurate results.

88h High-Throughput Screening of Binary and Ternary Organic Redox Active Materials-Based Deep Eutectic Solvents

Maria Politi, Jaime Rodriguez Jr., Stuart Adler, David Beck, Lilo Pozzo

Deep eutectic solvents (DESs) are a class of materials with varied applications including catalysis, synthesis, extraction processes, drug solubilization, and battery electrolytes. Their appeal stems from their broad electrochemical stability window, high electrical conductivity, low vapor-pressure, and low-flammability. These solvents possess a depression in the melting point at specific molar ratios of organic components that can result in a liquid solution near ambient temperatures. A breadth of candidates with varying concentrations can be used to form DESs, leading to a vast design space. To tackle such a challenge, we have demonstrated the use of high throughput experiments and data-driven design strategies for automated synthesis, rapid evaluation and screening of both their physical and electrochemical properties, for a total of 50 new combinations of DESs and DES analogous samples.[1] The implemented workflow uses low-budget systems (< $10k), commercially available materials and follows open-science principles. The current project aims to further expand the chemical space investigated using the high-throughput protocol for the formulation of binary and ternary deep eutectic solvents containing organic redox active materials (ORAM) for use as electrolytes in redox flow batteries (RFB). The use of bio-derived materials aims to be an alternative to expensive and resource-limited transition metals, such as the current industry leading RFB system based on Vanadium. The targeted organic redox compounds are selected from quinone derivatives (e.g., benzo-, anthra-quinones), fluorenones, and TEMPO. Once again, the physical and electrochemical properties of all synthesized samples have been screened following the high-throughput protocol demonstrated previously. Finally, to adapt high-throughput principles to the analysis of the data collection, machine learning was leveraged for gaining better understanding on the relationship between properties and structure of the starting materials and those of the final DES and DES analogous samples produced. The combination of high-throughput experimentation and data analysis greatly accelerated the design and screening of candidate binary and ternary organic redox active materials-based DES. This overall workflow could be easily adapted to other design spaces and applications.

416a Global Systems Analysis Using Deep Learning on Industrially Relevant Large Datasets

Gurkan Sin, Adem Rosenkvist Nielsen Aouichaoui

Global sensitivity analysis (GSA) is a thriving field of applied statistics that is used in the assessment of complex simulation models. Among its other uses, GSA is widely employed to generate insights into the contributions of individual model inputs, or sub-groups of inputs, to the variations in the output of a mechanistic model (Saltelli et al., 2019). However, when such models fall short of explaining a particular process phenomenon, sensitivity indices derived from these models become unreliable. One good example is nitrous-oxide emissions (N2O) from wastewater treatment plants (WWTP), for which the mechanistic understanding is still in its infancy(Daelman et al., 2015;Sin and Al., 2021). The recent breakthroughs achieved in deep learning (DL), on the other hand, offer an exciting possibility to bring new light to such poorly understood process phenomena. To this end, here we present a new framework, named deepGSA, incorporating well-established variance-decomposition-based global sensitivity analysis methods, such as Sobol sensitivity indices, with the plant data-driven deep learning modeling techniques.

The deepGSA aims at enabling non-specialist practitioners to leverage deep learning/machine learning models for GSA application purposes. To this end, the tool builds on an earlier GSA framework of the authors, easyGSA (Al et al., 2019), and is based on a recently proposed framework for DL-based and big data-driven process modeling (Hwangbo et al., 2020). By using these two frameworks, the deepGSA streamlines a number of tasks into a deep learning pipeline, such as data cleaning and preparation, model building and discrimination, model validation, Monte Carlo simulations, Sobol sensitivity analysis, Derivate-based global sensitivity analysis and effective visualizations of GSA results. The capabilities of the tool are highlighted with a case study from WWTPs concerning study of nitrogendioxide (N2O) emissions which is a potent greenhouse gas. For that purpose, a one-year-long dataset collected from four of the biological reactors of the Avedøre WWTP of Copenhagen (Denmark) (Chen et al., 2019) was used to train DL-based models. By using the tool, a number of candidate DL network topologies were evaluated, and a DL-model with satisfactory predictive performance (R^2_{test}>0.90) was obtained. By using this model in a parallelized Monte Carlo simulations procedure, the Sobol sensitivity indices were calculated to identify underlying factors (process disturbances and conditions) related to the emissions of N2O. Several sensitivity analysis techniques were applied to understand and explain which inputs are driving the greenhouse gas emissions. As regards the variance decomposition methods, the resutls indicate that Sobol total sensitivity index (STi) is an appropriate metrics to compare input/factor importance. Furthermore accounting for dependency among inputs, only influences the main effects (Si) calculations and not the STi. Hence STi is recommended a robust measure for relative importance of the inputs for this particular study. One disadvantage of variance based decomposition methods is that they do not indicate the direction of the effects of the inputs on the outputs (namely positive or negative influence) as Si and STi both are a ratio of the conditional variance of input on the output over total variance of outputs.

Derivative based sensitivity analysis also known as local elementary effects, in reference to Morris elementary effects, can reveal the sign of contributions (negative versus positive) of the inputs on the model outputs which provides a valuable information in engineering/scientific studies. When performed in a global setting (e.g. Sobol, I. M., andamp; Kucherenko, S. (2010)), properties of the distribution of these effects can be used to study and rank importance of factors. In this study, the distribution of the values of the said derivative functions revealed a pattern characterized by signifcantly heavy/fat tails in both negative and positive scales of function values. The normal distribution failed to described the tails, which is often used assumption in the literature. And this presents an interesting challenge as the standard inference statistics can not be used to interpret these results. To do a proper interpretation of the results, we turn to extreme value theory and statistics which studies the tailed distributions. Here the question addressed is how many monte carlo simulations (N) are needed to ensure convergence of such methods? To address these questions M/S plots from applied statistics to study the existence of moments of distributions are studied and discussed. We also study the pareto distribution funciton and its estimation of the tail index using hill estimators. All these analysis indicate/confirm the presence of a heavy tailed distribution. We conclude with a new methodology for performing a global systems analysis of industrial datasets particularly subject to fat tailed distribution of their inputs.

References

Al, R., Behera, C.R., Zubov, A., Gernaey, K. V., Sin, G., 2019. Meta-modeling based efficient global sensitivity analysis for wastewater treatment plants – An application to the BSM2 model. Comput. Chem. Eng. 127, 233–246. https://doi.org/10.1016/j.compchemeng.2019.05.015

Chen, X., Mielczarek, A.T., Habicht, K., Andersen, M.H., Thornberg, D., Sin, G., 2019. Assessment of Full-Scale N 2 O Emission Characteristics and Testing of Control Concepts in an Activated Sludge Wastewater Treatment Plant with Alternating Aerobic and Anoxic Phases. Environ. Sci. Technol. 53, 12485–12494. https://doi.org/10.1021/acs.est.9b04889

Daelman, M.R.J., van Voorthuizen, E.M., van Dongen, U.G.J.M., Volcke, E.I.P., van Loosdrecht, M.C.M., 2015. Seasonal and diurnal variability of N 2 O emissions from a full-scale municipal wastewater treatment plant. Sci. Total Environ. 536, 1–11. https://doi.org/10.1016/j.scitotenv.2015.06.122

Hwangbo, S., Al, R., Sin, G., 2020. An integrated framework for plant data-driven process modeling using deep-learning with Monte-Carlo simulations. Comput. Chem. Eng. 143, 107071. https://doi.org/10.1016/j.compchemeng.2020.107071

Saltelli, A., Aleksankina, K., Becker, W., Fennell, P., Ferretti, F., Holst, N., Li, S., Wu, Q., 2019. Why so many published sensitivity analyses are false: A systematic review of sensitivity analysis practices. Environ. Model. Softw. 114, 29–39. https://doi.org/10.1016/j.envsoft.2019.01.012

Sin, G., andamp; Al, R. (2021). Activated sludge models at the crossroad of artificial intelligence—A perspective on advancing process modeling. *npj Clean Water*, *4*(1), 1-7.

416b Uncertainty Quantification for Molecular Property Predictions Using Automatic Graph Neural Architecture Search

Shengli Jiang, Shiyi Qin, Prasanna Balaprakash, Reid Van Lehn, Victor Zavala

Generating and evaluating candidates for new molecules using data-driven models is essential in modern synthesis platforms. Quantitative structure–activity relationships (QSARs) provide a powerful data-driven modeling framework for calculating molecular properties using expensive and time-consuming experiments but they tend to be limited in predictability and in that the user needs to pre-define structural features [1]. Over the past few years, the fields of molecular property prediction have been advanced by using deep learning techniques, especially neural networks (NNs) [2]. NNs can leverage flexible representations of molecules that can be tailored to tasks rather than depending on expert-defined molecular features [3]–[5]. However, despite their progress, the use of NNs for molecular modeling still faces a number of limitations. Specifically, the lack of expressiveness and transparency of NNs makes it difficult to assess their robustness, out-of-domain applicability, and potential failure modes. This is particularly relevant when seeking to embed NNs in design of experiments tasks. To overcome these limitations, uncertainty quantification (UQ) capabilities need to be provided [6]. UQ is a set of mathematical techniques that aims to characterize data uncertainty and model uncertainty; the former occurs as a result of the inherent variability or noise in the data, while the latter is associated with the model parameter estimation or out-of-distribution predictions [7]. While data uncertainty is typically assumed to be irreducible, model uncertainty can be reduced by gathering more training data in appropriate regions of an experimental space [8].

Recent work has been done in UQ for NN models within the context of molecular prediction [9]–[14]. Recently, Hirschfeld et al. systematically evaluated a variety of UQ techniques on message-passing neural nets (MPNNs), which learn parameterized mappings from graph-structured objects to feature vectors and have achieved state-of-the-art performance across various industrial datasets. Probabilistic models, such as full Bayesian formulations, can quantify uncertainty but is computationally intractable for MPNN models that contain millions of trainable parameters. Ensemble approaches that use multiple independently trained MPNNs have shown promise in terms of scalability [15]. Here, each candidate model in the ensemble can be trained in parallel, drastically reducing the computational time. A critical component of the ensemble is its diversity, without which uncertainty cannot be properly quantified. For example, models with the same NN architecture but different weight initializations can result in poor estimates of model uncertainty [16]. The lack of scalable UQ capabilities in NN models limits the use of active learning and experimental design.

In this work, we propose an automated approach to construct diverse MPNN models using an adaption of the AutoDEUQ method [16]. Here, we used aging evolution to find candidate models automatically. The aging evolution (AE) method constructs the initial population of MPNNs by sampling a set N of random architectures. It evaluates the initial population and record the validation loss from each individual. To quantify uncertainty, the approach uses the negative log-likelihood loss (as opposed to the usual mean squared error) in the training. Following the initialization, AE samples S random architectures uniformly from the population with replacement. The architecture with the lowest validation loss within the sample is selected as a parent. A mutation is performed on the parent, and new child architecture is constructed. A mutation corresponds to choosing a different MPNN layer operation (e.g., activation function, number of hidden units). The child is trained, and the validation loss is recorded. Consequently, the child is added to the population by replacing the oldest architecture in the population. Over multiple cycles, architectures with lower validation loss are retained in the population via repeated sampling and mutation. The next step is to select the top-k

MPNNs from the search to build the ensemble. We leveraged variance decomposition to separate the data and model uncertainty from the predicted variance of the ensemble. The data uncertainty is the mean of the predicted variance of the ensemble, and the model uncertainty is the variance of the predicted mean. We used the negative log-likelihood as the UQ metric and demonstrated an improved UQ compared to previous ensemble methods on various benchmark datasets, including QM7 and ESOL. We also extended our UQ method to a new multi-molecular dataset for activity coefficient estimation and demonstrated promising results.

References

[1] A. Cherkasov et al., "QSAR modeling: Where have you been? Where are you going to?," Journal of Medicinal Chemistry, vol. 57, no. 12. 2014. doi: 10.1021/jm4004285.

[2] W. P. Walters and R. Barzilay, "Applications of Deep Learning in Molecule Generation and Molecular Property Prediction," Accounts of Chemical Research, vol. 54, no. 2, 2021, doi: 10.1021/acs.accounts.0c00699.

[3] E. N. Feinberg et al., "PotentialNet for Molecular Property Prediction," ACS Central Science, vol. 4, no. 11, 2018, doi: 10.1021/acscentsci.8b00507.

[4] Z. Hao et al., "ASGN: An Active Semi-supervised Graph Neural Network for Molecular Property Prediction," 2020. doi: 10.1145/3394486.3403117.

[5] J. Gilmer, S. S. Schoenholz, P. F. Riley, O. Vinyals, and G. E. Dahl, "Neural message passing for quantum chemistry," in 34th International Conference on Machine Learning, ICML 2017, 2017, vol. 3.

[6] Y. Gal and Z. Ghahramani, "Dropout as a Bayesian approximation: Representing model uncertainty in deep learning," in 33rd International Conference on Machine Learning, ICML 2016, 2016, vol. 3.

[7] E. Hüllermeier and W. Waegeman, "Aleatoric and epistemic uncertainty in machine learning: an introduction to concepts and methods," Machine Learning, vol. 110, no. 3, 2021, doi: 10.1007/s10994-021-05946-3.

[8] Y. Gal and Z. Ghahramani, "Dropout as a Bayesian approximation: Representing model uncertainty in deep learning," in 33rd International Conference on Machine Learning, ICML 2016, 2016, vol. 3.

[9] J. P. Janet, C. Duan, T. Yang, A. Nandy, and H. J. Kulik, "A quantitative uncertainty metric controls error in neural network-driven chemical discovery," Chemical Science, vol. 10, no. 34, 2019, doi: 10.1039/c9sc02298h.

[10] N. Aniceto, A. A. Freitas, A. Bender, and T. Ghafourian, "A novel applicability domain technique for mapping predictive reliability across the chemical space of a QSAR: Reliability-density neighbourhood," Journal of Cheminformatics, vol. 8, no. 1, 2016, doi: 10.1186/s13321-016-0182-y.

[11] R. Liu and A. Wallqvist, "Molecular Similarity-Based Domain Applicability Metric Efficiently Identifies Out-of-Domain Compounds," Journal of Chemical Information and Modeling, vol. 59, no. 1, 2019, doi: 10.1021/acs.jcim.8b00597.

[12] Y. Zhang and A. A. Lee, "Bayesian semi-supervised learning for uncertainty-calibrated

prediction of molecular properties and active learning," Chemical Science, vol. 10, no. 35, 2019, doi: 10.1039/c9sc00616h.

[13] S. Ryu, Y. Kwon, and W. Y. Kim, "Uncertainty quantification of molecular property prediction with Bayesian neural networks," Mar. 2019, doi: 10.48550/arxiv.1903.08375.

[14] K. Tran, W. Neiswanger, J. Yoon, Q. Zhang, E. Xing, and Z. W. Ulissi, "Methods for comparing uncertainty quantifications for material property predictions," Machine Learning: Science and Technology, vol. 1, no. 2, 2020, doi: 10.1088/2632-2153/ab7e1a.

[15] L. Hirschfeld, K. Swanson, K. Yang, R. Barzilay, and C. W. Coley, "Uncertainty Quantification Using Neural Networks for Molecular Property Prediction," Journal of Chemical Information and Modeling, vol. 60, no. 8, 2020, doi: 10.1021/acs.jcim.0c00502.

[16] R. Egele, R. Maulik, K. Raghavan, B. Lusch, I. Guyon, and P. Balaprakash, "AutoDEUQ: Automated Deep Ensemble with Uncertainty Quantification," Oct. 2021, Accessed: Mar. 04, 2022. [Online]. Available: https://arxiv.org/abs/2110.13511v2

416c Projecting the Effectiveness of Deep Ensembles

Charles McGill, William Green

Deep ensembling is a core practice in the creation of robust machine learning models, improving prediction performance through the combination of independent submodels. In this work, we explore the practice of ensembling in chemical systems statistically in order to quantify the contribution of model variance error, evaluate the use of ensemble metrics to estimate model uncertainty, and project the performance of models using different ensemble sizes. This ability to project ensemble performance addresses the key tension of ensemble training: the diminishing returns for increasingly large ensembles and the related uncertainty of how many ensemble submodels to train.

Here we introduce a method of Bayesian inference to characterize the distribution of potential model predictions from which we sample during ensembling. This approach is completely posterior to the practice of training the model, requiring no alteration of the original model architecture. Through characterization of this distribution, we are able to separate he portion of model prediction errors that are due to model variance from those that would still be present with an infinite number of ensemble models. We are able to use this separation to evaluate which model regimes are subject to errors correctable by ensembling and those that are not. Further, we are able to use this distribution to estimate with good accuracy the expected performance of a model with a larger number of ensemble submodels. We demonstrate the robustness of these projections in common benchmark datasets as well as an artificially constructed dataset where the level of data noise errors can be controlled accordingly. We also evaluate the cases in which the variance of ensemble predictions can be useful as a metric of uncertainty and show the limitations of this metric to predict nonvariance errors when calibrated or uncalibrated.

416d Automatic Strain Sensor Design *Via* Active Learning and Data Argumentation for Soft Machines

Po-Yen Chen, Tianle Chen

Machine learning (ML) is a versatile tool that builds up a model to make predictions or recommendations, which has been particularly appropriate for the data-rich systems to uncover underlying trends and investigate complex correlations. The exploitation of machine intelligence to research in materials science has gained many interests in the subjects of organic/inorganic catalyst design, drug discovery, and quantum dot synthesis, where a large number of data points can be supplied either computationally or analytically. In contrast, at the device level, the data acquisition rate is primarily limited by the scarcity of high-quality data. For example, substantial challenges exist in developing a prediction model with high accuracy to design piezoresistive strain sensors. First, there are no standardized protocols for the fabrication/characterization of soft strain sensors, and different research laboratories come up with their own fabrication schemes. Thus, the sensor data collected from the literature are with poor consistency and may contain bias. Second, most of recent reports focused on the fabrication of strain sensors enabling real-time monitoring of in-plane strains between 30% to 200%. Such unbalanced data is likely to produce a ML model with restricted prediction capabilities within specific strain ranges, which would not benefit the strain sensor design for many soft machines at extreme strain cases (e.g., < 5% or > 300%). Last, the fabrication of soft strain sensors is normally tedious and time-consuming, so the data acquisition rate is low, making the ML model fall short of data points and present insufficient prediction accuracy.
As a result, it is of high value to address these challenges through a hybrid approach (experimentally and computationally) to construct a ML model at device level, which can realize the concept of automatic strain sensor design for diverse soft machines. In this presentation, a hybrid approach through wet-lab experiments and machine intelligence was developed to automate the design of strain sensors across a wide strain range (from <0.5% to 350%) by developing a high-accuracy prediction model. A three-stage ML framework, including boundary definition, active learning, and data augmentation, was realized to cultivate the prediction ML model with high accuracy. To ensure data quality and balance, 351 delamination tests and 125 nanomaterials-based strain sensors were conducted/fabricated by one lab under identical protocols. After data augmentation and genetic algorithm-based selection, the multi-dimensional dataset of ML-enabled prediction model contained >10,000 real and virtual data points. An ultimate prediction model was finally constructed and able to (1) predict sensor characteristics based on fabrication recipes and (2) recommend feasible fabrication recipes for adequate strain sensors. As final demonstrations, model-suggested strain sensors were integrated into soft gripper and batoid-like swimmer to endow them with real-time sensing capabilities.

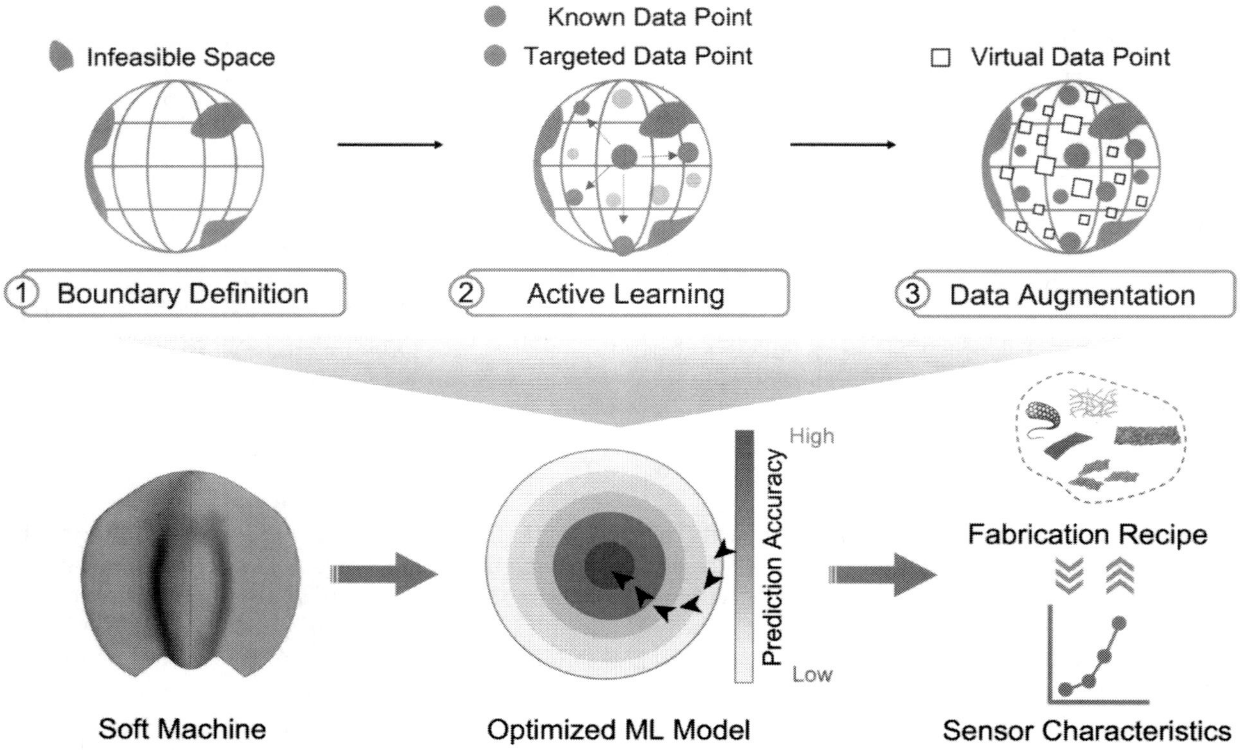
Infeasible Space
Known Data Point
Targeted Data Point
Virtual Data Point
① Boundary Definition
② Active Learning
③ Data Augmentation
High
Prediction Accuracy
Low
Fabrication Recipe
Soft Machine
Optimized ML Model
Sensor Characteristics

416e A Framework for the Statistical Value of Extreme Points for Function Approximation

Arijit Chakraborty, Krishnan Kumaran, Venkat Venkatasubramanian

With the recent data deluge in various science and engineering disciplines, there is renewed interest in data-driven techniques for function approximation. A necessary component of such a problem is the availability of data that is representative of the input space such that it can be used to reasonably approximate the true function. The statistical framework of experimental design allows one to estimate the relative value of data-points, subject to optimization of a user-defined objective function. Such problems requiring optimal design points frequently arise in kinetics of complex reaction systems, developmental toxicology, and thermodynamic calculations of activity coefficients for mixtures – among other applications. We argue that not all data-points are equally beneficial, nor are they all necessary. We posit that there are data regimes in the function's domain which provide more information than others. Using model-based experimental design, the case in favor of extreme points is presented, along with their statistical merits. We demonstrate that this results in the lowest value of parameter uncertainty. Additionally, we formulate an optimization problem that corroborates our preference of extreme points of a function in comparison to any other data-points in the input domain. The methodology can be extended to higher dimensions, and we present how such an approach can be utilized by an end-user to reduce time required for performing experiments.

416f Graph Hysteria – Comparing the Generative Performance of Graph and String-Based Translation Vaes for Molecular Design

Orion Dollar, Nisarg Joshi, David Beck, Jim Pfaendtner

Graph representations have grown steadily more popular as input representations for *de novo* molecular design models.[1–3] The inherent coherence between the nodes and edges of a graph and the atoms and bonds within a molecule lend graph-based architectures a natural compatibility with tasks requiring a machine-learned representation of molecular structure. The flexibility of graphs also allows for the incorporation of additional physicochemical features which have been shown to be very effective at improving property prediction models.[4] However, their increased complexity compared to 1D string representations imbues them with some undesirable properties such as reduced computational efficiency[5] and reduced performance on standard generative benchmarks.[6]

Here we systematically compare the generative and predictive properties of graph and string-based encoders and decoders by framing the task of a VAE as a set of machine-translation problems – graph-to-graph, string-to-string, graph-to-string and string-to-graph. In doing so we can isolate the impact of the input representation on the quality of the learned molecular embeddings as well as the impact of the output representation on the novelty, diversity and validity of machine-generated structures. We find that the choice of encoder has a tangible effect on the model's ability to explore molecular phase space and that the choice of decoder has significant influence on the practical viability of a model. Finally, we also compare the effect of input representation on property prediction and model interpretability and discuss in which scenarios each architecture is likely to be optimal.

1. Jin, W., Barzilay, R. andamp; Jaakkola, T. Junction Tree Variational Autoencoder for Molecular Graph Generation. *arXiv:1802.04364 [cs.LG]* (2018).
2. Jin, W., Barzilay, R. andamp; Jaakkola, T. Hierarchical Generation of Molecular Graphs using Structural Motifs. in *Proceedings of the 37th International Conference on Machine Learning* 4839–4848 (2020).
3. Mahmood, O., Mansimov, E., Bonneau, R. andamp; Cho, K. Masked graph modeling for molecule generation. *Nature Communications 2021 12:1* **12**, 1–12 (2021).
4. Gasteiger, J., Groß, J. andamp; Günnemann, S. Directional Message Passing for Molecular Graphs. *arXiv:2003.03123* (2020) doi:10.48550/arxiv.2003.03123.
5. Mercado, R. *et al.* Graph networks for molecular design. *Machine Learning: Science and Technology* **2**, 025023 (2021).
6. Flam-Shepherd, D., Zhu, K. andamp; Aspuru-Guzik, A. Keeping it Simple: Language Models can learn Complex Molecular Distributions. *arXiv:2112.03041* (2021) doi:10.48550/arxiv.2112.03041.

416g Extending Bigsmiles to Non-Covalent Bonds in Supramolecular Polymer Assemblies

Weizhong Zou, Alexis Martell-Monterroza, Yunxin Yao, S. Cem Milik, Morgan Cencer, Nathan Rebello, Haley Beech, Melody Morris, Tzyy-Shyang Lin, Cleotilde Castano, Julia Kalow, Stephen Craig, Alshakim Nelson, Jeffrey Moore, Bradley Olsen

As a machine-recognizable representation of polymer connectivity, BigSMILES line notation extends SMILES from deterministic to stochastic structures. The same framework that allows BigSMILES to accommodate stochastic covalent connectivity can be extended to non-covalent bonds, enhancing its value for polymers, supramolecular materials, and colloidal chemistry. Non-covalent bonds are captured through the inclusion of annotations to pseudo atoms serving as complementary binding pairs, minimal key/value pairs to elaborate other relevant attributes, and indexes to specify the pairing among potential donors and acceptors or bond delocalization. Incorporating these annotations into BigSMILES line notation enables the representation of four common classes of non-covalent bonds in polymer science: electrostatic interactions, hydrogen bonding, metal-ligand complexation, and π-π stacking. The principal advantage of non-covalent BigSMILES is the ability to accommodate a broad variety of non-covalent chemistry with a simple user-orientated, semi-flexible annotation formalism. This goal is achieved by encoding a universal but non-exhaustive representation of non-covalent or stochastic bonding patterns through syntax for (de)protonated and delocalized state of bonding as well as nested bonds for correlated bonding and multi-component mixture. By allowing user-defined descriptors in the annotation expression, further applications in data-driven research can be envisioned to represent chemical structures in many other fields, including polymer nanocomposite and surface chemistry.

416h Formal Proofs in the Chemical Sciences Using the Lean Theorem Prover

Parivash Feyzishendi, Maxwell P. Bobbin, Catherine M. Wraback, Samiha Sharlin, Tyler R. Josephson

In contrast to derivations that are simply hand-written or typeset in a research article, a formalized proof is a computer-verified sequence of steps logically relating the axioms of a problem to a conjecture to be proved.

Mathematicians have formalized tens of thousands of proofs in self-consistent libraries[1], but formal theorem proving has received less attention in science and engineering.

Formalizing proofs in the chemical sciences is worthwhile for at least two reasons. Firstly, formal proofs enable more rigorous science. Formalized proofs are guaranteed to be valid, and require every assumption to be laid out in transparent, mathematical precision. Scientists can focus on the assumptions and definitions, trusting that any proved conjectures necessarily follow. Secondly, formal proofs are machine-readable and interpretable. This is enabling automated problem solving in pure mathematics, for example, artificial intelligence tools have solved problems from the International Math Olympiad[2].

To formalize proofs in the chemical sciences, we propose to use the Lean language and theorem prover[1]. It can express the advanced math required in our most elegant proofs (e.g. statistical, quantum, and fluid mechanics), which sometimes require (random variables, partial differential equations, and N-dimensional integrals).

We show how a few prototypical proofs can be developed in Lean, including the kinetic derivation of the Langmuir adsorption model[3], as well as the BET theory of multilayer adsorption, which invokes two infinite geometric series in its proof[4]. By checking the conditions of each lemma used in the proof, Lean guarantees that the proof is valid and consistent with all the other math in its database.

1. L. de Moura, et al.. The Lean Theorem Prover. 25th International Conference on Automated Deduction (CADE-25), Berlin, Germany, 2015
2. S. Polu, et al.. Formal Mathematics Statement Curriculum Learning, arXiv:2202.01344 [cs], 2022
3. I. Langmuir, THE ADSORPTION OF GASES ON PLANE SURFACES OF GLASS, MICA AND PLATINUM, JACS 40, 1361, 1918
4. S. Brunauer, P. H. Emmett, E. Teller. Adsorption of Gases in Multimolecular Layers, JACS 60, 309, 1938

416i Machine Learning with Differential Equation Priors

Aditi Krishnapriyan

Modeling the differential equations governing physical systems is a central aspect of many science and engineering tasks. These systems verify exact constraints corresponding to the laws of physics. Recently, neural network approaches have shown promise in solving differential equations via approximating functions. However, these approaches only approximately enforce the constraints on a physical system. I will first discuss the challenges associated with such an approach. I will then discuss overcoming these challenges by developing a neural network architecture that incorporates differential equation constrained optimization, which outputs solutions that verify the desired physical constraints exactly over a given spatial and/or temporal domain. I will show that this architecture allows us to accurately and efficiently fit solutions to new problems, and demonstrate this on fluid flow and transport phenomena problems.

218a Discovery of O_2-Selective Metal-Organic Frameworks Via Bayesian Optimization

Eric Taw, Jeffrey B. Neaton

Metal-organic frameworks (MOFs) present opportunities to tune chemical selectivity towards O2 by changing metal species and organic linkers. However, this also presents a combinatorial number of possible MOF structures, making a high-throughput screening study impossible with expensive quantum chemistry methods. We have shown previously that a machine learning technique called Bayesian optimization substantially reduces the number of calculations needed to find the top candidate MOF, thus enabling the screening of databases with time-consuming calculations. [1]

Here, we apply this technique to discover new O2-selective air separation MOFs using accurate density functional theory (DFT) calculations. Starting with the CoRE MOF database [2] and the enthalpy of adsorption of O2 in various MOFs in the experimental literature [3], we allow Bayesian optimization to suggest MOFs in the database to evaluate with DFT. We derive a new acquisition function, denoted as mean-squared-error expected improvement (MSE-EI) that allows us to target a binding enthalpy of 45 kJ/mol. This corresponds to roughly half the energy consumption needed for a state-of-the-art pressure swing adsorption system for air separation. We present the MOF candidates found in this procedure for further experimental validation.

[1] Taw, E.; Neaton, J.B.; Adv. Theory and Sim. 2022, 5, 3, 2100515
[2] Chung, Y.G. et al.; J. Chem. Eng. Data 2019, 64, 12, 5985-5998
[3] Jaramillo, D.; ... Taw, E. et al; in preparation

218c Cript: A Scalable Polymer Material Data Structure

Dylan Walsh, Bradley Olsen, Klavs Jensen

Having accessible well-structured data is the foundation of cheminformatics. The complexity of polymer structures poses significant challenges in the formation of databases as there is no single representation that can capture the full molecular detail of a polymer material. More specifically, polymers are large stochastic molecules with distributions in chain length, composition, and topology. Additionally, data collection methods are highly variable, typically provide relative structural information (ex. molecular weight relative to a polystyrene standard), and/or use models which require expert knowledge to put into context. In some cases, experimentally obtaining structural information is impossible, and information from prior processing steps is needed. To complicate matters further, polymers can assemble into a wide range of morphologies through phenomena like phase segregation and crystallization. The formation of these morphologies can be highly influenced by the processing condition under which the material was made. Ultimately, data sets that do not completely capture all the relevant polymer, process, and characterization information pose challenges for advancing data-driven research in the polymer field.

Here, we discuss the development and deployment of CRIPT to the community. CRIPT is an open-source community driven digital polymer data ecosystem. At the core of CRIPT is universal data model that provides a unique graph-based representation which enables the full scope of polymer data to be captured and organized in an intuitive manner. This includes experimental data such as synthesis, material processing, chemical characterization, and physical characterization, as well as atomistic simulation data. To support the data model, we have launched a website and python API to enable everyone, from chemists to data scientists, to add and retrieve data in an efficient manner. This seamless access to polymer data will enable solutions which otherwise go unseen to be a simple search away which will lead to the ability to develop better materials faster. Overall, CRIPT seeks to accelerate the development of new innovations and enable the rapid sharing of data across research teams.

218d Data-Driven Design of Polymer-Protein Hybrid Materials

Michael Webb

Polymer-protein hybrids are intriguing materials that can bolster protein stability in non-native environments, thereby enhancing their utility in diverse medicinal, commercial, and industrial applications. One stabilization strategy involves designing synthetic random copolymers with compositions attuned to the protein surface, but rational design is complicated by a vast chemical and composition space. In this talk, I will describe our strategy to design protein-stabilizing copolymers based on active machine learning, facilitated by automated material synthesis and characterization platforms. The versatility and robustness of the approach is demonstrated by the successful identification of copolymers that preserve, or even enhance, the activity of chemically distinct enzymes following exposure to thermal denaturing conditions. Using post hoc analysis, molecular modeling, and additional experimental characterization, I will discuss likely reasons underlying the success of these polymer-protein combinations. Overall, this demonstrates a useful application of data science to an otherwise complex design problem, enabling the study of interesting systems that we would not be in a position to identify relying on more conventional methods.

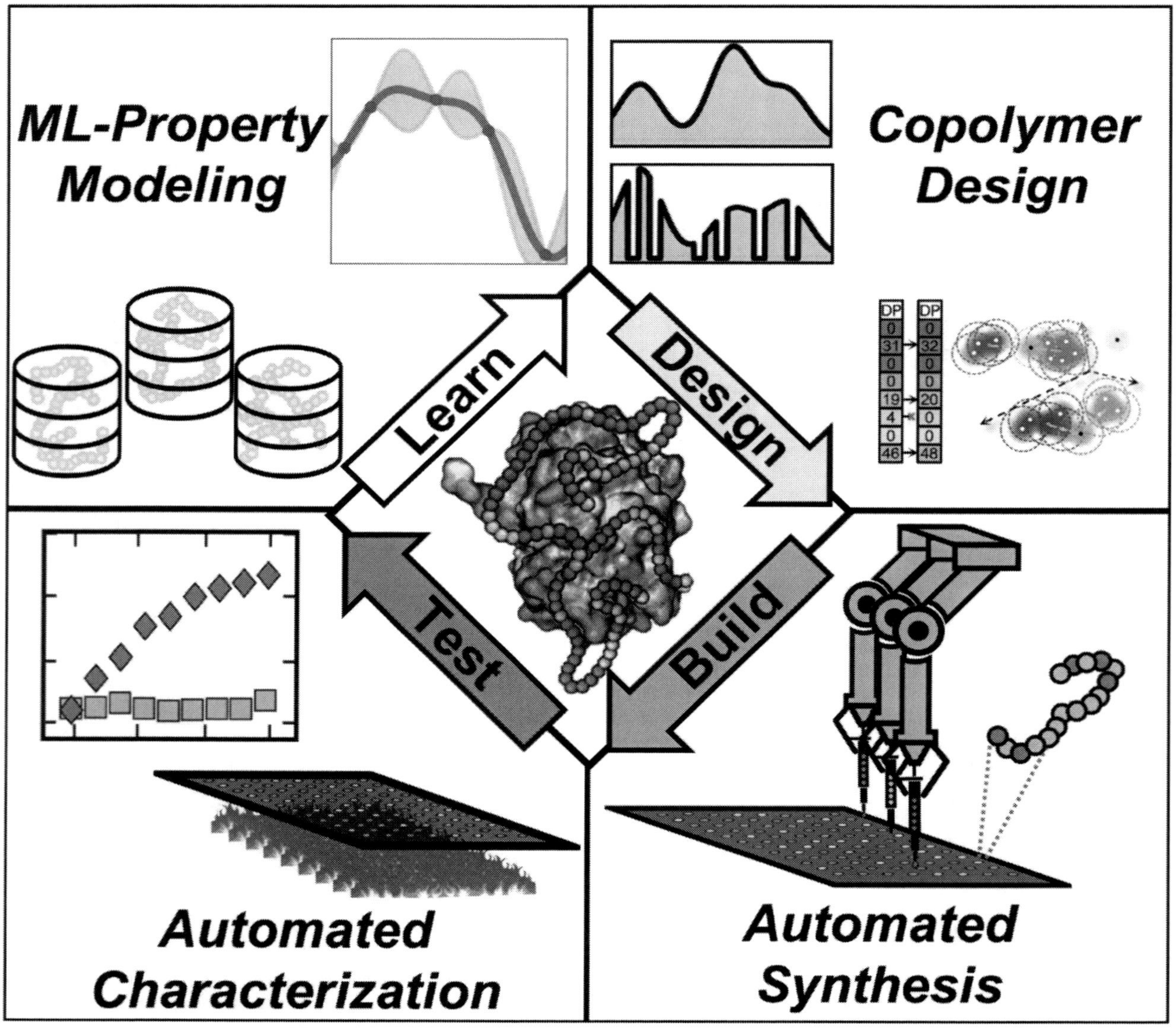

218e Applications of Machine-Learned Electron Densities of Nucleic Acids

Alex Lee, Joshua A. Rackers, William P. Bricker

The efficient prediction of large-scale electron densities of biological macromolecules and molecular arrays could lead to breakthroughs in quantum chemistry, structural biology, and molecular design. This work leverages a novel machine learning (ML) algorithm termed Euclidean Neural Networks to learn the electron densities of arbitrary sequences of DNA by performing many smaller *ab initio* calculations of the component base-pair steps of DNA. These electron density predictions are currently accurate to around 1% error for arbitrary sequences of DNA, and are easily size-extensible unlike traditional quantum calculations (Lee et al. 2022). Using this machine learning algorithm, electron density predictions of large-scale DNA structures consisting of tens-of-thousands to hundreds-of-thousands of electrons can be performed in minutes. Several applications of these large-scale machine-learned electron densities are of interest. We focus here on the accurate calculation of energies and forces from the machine-learned electron densities, which allows for dynamical propagation of the electronic system, termed *ab initio* molecular dynamics (AIMD). These machine-learned electron densities would allow for AIMD simulations of systems that are much larger than can be propagated using traditional quantum calculations. In addition, these forces can be utilized to develop more accurate force fields for classical molecular dynamics. Other applications such as the use of machine-learned electron densities to aid in X-ray crystal refinement of large biomolecular structures will be discussed in brief.

AJ Lee, JA Rackers, WP Bricker. Predicting quantum-accurate electron densities for DNA with equivariant neural networks. *ChemRxiv* (10 Mar 2022).

218f Using Machine Learning Approaches to Estimate Enzyme Kinetic Parameters

Veda Sheersh Boorla, Costas D. Maranas

The catalytic turnover number (k_{cat}) is a key kinetic property an enzyme defined as the maximal number of molecules of substrates converted to products per active site per unit time. Accurate and generalizable methods for estimating enzyme kinetic parameters such as the one presented here can be invaluable for applications ranging from metabolic modeling to enzyme re-engineering. Enzyme databases such as BRENDA[1] contain a repository of turnover numbers measured *in vitro*. However, the available data is noisy due to various experimental conditions and lack of proper annotations for several entries. We curated a dataset of ~6,000 turnover numbers from BRENDA by applying several quality filters. Using this dataset, we trained a convolutional neural network (CNN) model to learn amino-acid embeddings that accurately estimate k_{cat} values when used as enzyme features along with morgan fingerprints as substrate features. The trained model achieved an average Pearson correlation coefficient of **0.78** (standard deviation 0.008) and an average root mean squared error of **0.96** (standard deviation 0.061) in a 5-fold cross validation evaluation. The root mean squared error of 0.96 in log scale corresponds to less than an order of magnitude error in linear scale which is quite low compared to the overall range of k_{cat} values (1E-06 to 1E+07). The low standard deviations across cross-validation suggests a robust and generalizable training across the entire dataset. The success of our model can be attributed to the ability of CNNs to extract complex local patterns of amino acid residues that may be responsible for actual enzyme-substrate interactions on the molecular level. Comparison of our model to existing methods along with its current limitations and provision for improvements will be discussed. In particular, the use of state-of-the-art protein language model embeddings as features and the use of graph-based architectures that overcome the limitations and CNNs and provide more meaningful insights will be discussed. By training ML models to accurately capture the mapping between amino acid mutations and changes in turnover numbers, they can be used to guide directed evolution and/or targeted enzyme engineering approaches.

References:

1. Chang, Antje et al. "BRENDA, the ELIXIR core data resource in 2021: new developments and updates." Nucleic acids research vol. 49,D1 (2021): D498-D508.

218g Classifying the Toxicity of Pesticides to Honey Bees Via Support Vector Machines with Random Walk Graph Kernels

Ping Yang, Adrian Henle, Xiaoli Fern, Cory Simon

Pesticides benefit agriculture by increasing crop yield, quality, and security. However, pesticides may inadvertently harm bees, which are agriculturally and ecologically vital as pollinators. The development of new pesticides---driven by pest resistance to and demands to reduce negative environmental impacts of incumbent pesticides---necessitates assessments of pesticide toxicity to bees. We leverage a data set of 382 molecules labeled from honey bee toxicity experiments to train a classifier that predicts the toxicity of a new pesticide molecule to honey bees. Traditionally, the first step of a molecular machine learning task is to explicitly convert molecules into feature vector representations for input to the classifier. Instead, we (i) adopt the fixed-length random walk graph kernel to express the similarity between any two molecular graphs and (ii) use the kernel trick to train a support vector machine (SVM) to classify the bee toxicity of pesticides represented as molecular graphs. We assess the performance of the graph-kernel-SVM classifier under different walk lengths used to describe the molecular graphs. The optimal classifier, with walk length 5, achieves an (mean over 100 runs) accuracy, precision, and recall of 0.83, 0.71, and 0.72 on a test data set.

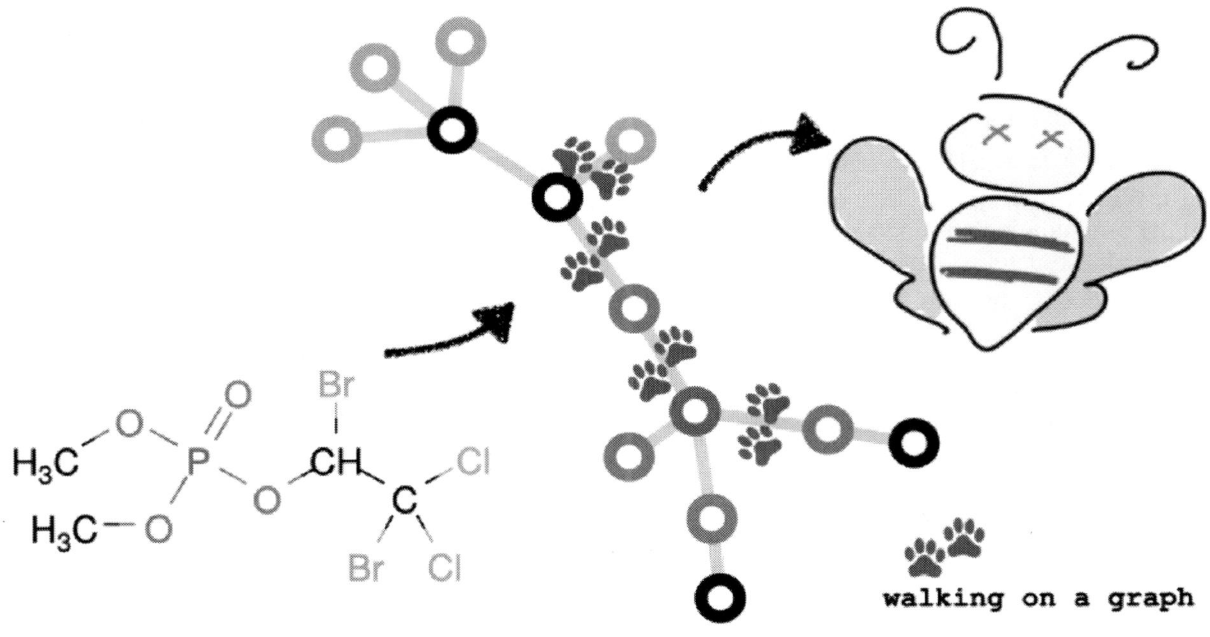

218h Hallucinating Inexpensive, Diverse and Native-like Antibody Binders with Deep Learning

Sai Pooja Mahajan, Jeffrey Ruffolo, Rahel Frick, Jeffrey J. Gray

Antibodies recognize and bind an extremely diverse repertoire of antigens via 6 hypervariable loop regions known as the complementarity determining regions or the CDRs. A common goal in antibody engineering to generate a library of CDRs enriched in binders also known as affinity maturation. Experimental methods for affinity maturation are expensive, laborious, and time-consuming and rarely allow the efficient exploration of the full design space. Conventional computational methods are relatively inexpensive but marred by inaccurate scoring functions and slow and inefficient sampling of the full combinatorial sequence space. Deep learning (DL) models are transforming the field of protein structure-prediction, engineering, and design.[1] While several DL-based protein design methods have shown promise, specialized models for antibody design are highly desirable.[2] Our approach aims to fully leverage structural information which is becoming increasingly abundant in the era of highly accurate structure prediction models such as AlphaFold[3] (all proteins, multimers) and DeepAb[4] (antibodies only). Inspired by the hallucination frameworks that specifically leverage such structure prediction DL models, we propose the F_vHallucinator for generating antibody sequences conditioned on a target antibody structure with DeepAb.[2,5,6] On a benchmark set of 60 antibodies, the F_vHallucinator recovers over 50% of the wildtype CDR sequence on all six CDR loops. At the V_H-V_L interface, the F_vHallucinator designs amino acid substitutions that are highly enriched in human repertoire sequences. Furthermore, when compared to a large experimentally characterized library of CDR H3s of the anti-HER2 antibody,[7] trastuzumab, the F_vHallucinator designs exhibit high sequence identity to known HER2-binders. We propose a pipeline that screens F_vHallucinator designs to obtain a virtually screened library enriched in binders for an antigen of interest. We apply this pipeline to the CDR H3 of the trastuzumab-HER2 complex to generate designs that retain the original binding mode and improve the binding affinity and interfacial properties. Thus, the F_vHallucinator pipeline enables fast and inexpensive generation of diverse, structure-conditioned antibody libraries enriched in binders.

1. Gao, W., Mahajan, S. P., Sulam, J. andamp; Gray, J. J. Deep Learning in Protein Structural Modeling and Design. *Patterns* **1**, 100142 (2020).
2. Norn, C. *et al.* Protein sequence design by explicit energy landscape optimization. doi:10.1101/2020.07.23.218917.
3. Jumper, J. *et al.* Highly accurate protein structure prediction with AlphaFold. *Nature* (2021) doi:10.1038/s41586-021-03819-2.
4. Ruffolo, J. A., Sulam, J. andamp; Gray, J. J. Antibody structure prediction using interpretable deep learning. *Patterns* **3**, 100406 (2022).
5. Anishchenko, I. *et al.* De novo protein design by deep network hallucination. *Nature* 1–19 (2020).
6. Wang, J. *et al.* Deep learning methods for designing proteins scaffolding functional sites. *bioRxiv* 2021.11.10.468128 (2021).
7. Mason, D. M. *et al.* Optimization of therapeutic antibodies by predicting antigen specificity from antibody sequence via deep learning. *Nat. Biomed. Eng.* (2021) doi:10.1038/s41551-021-00699-9.

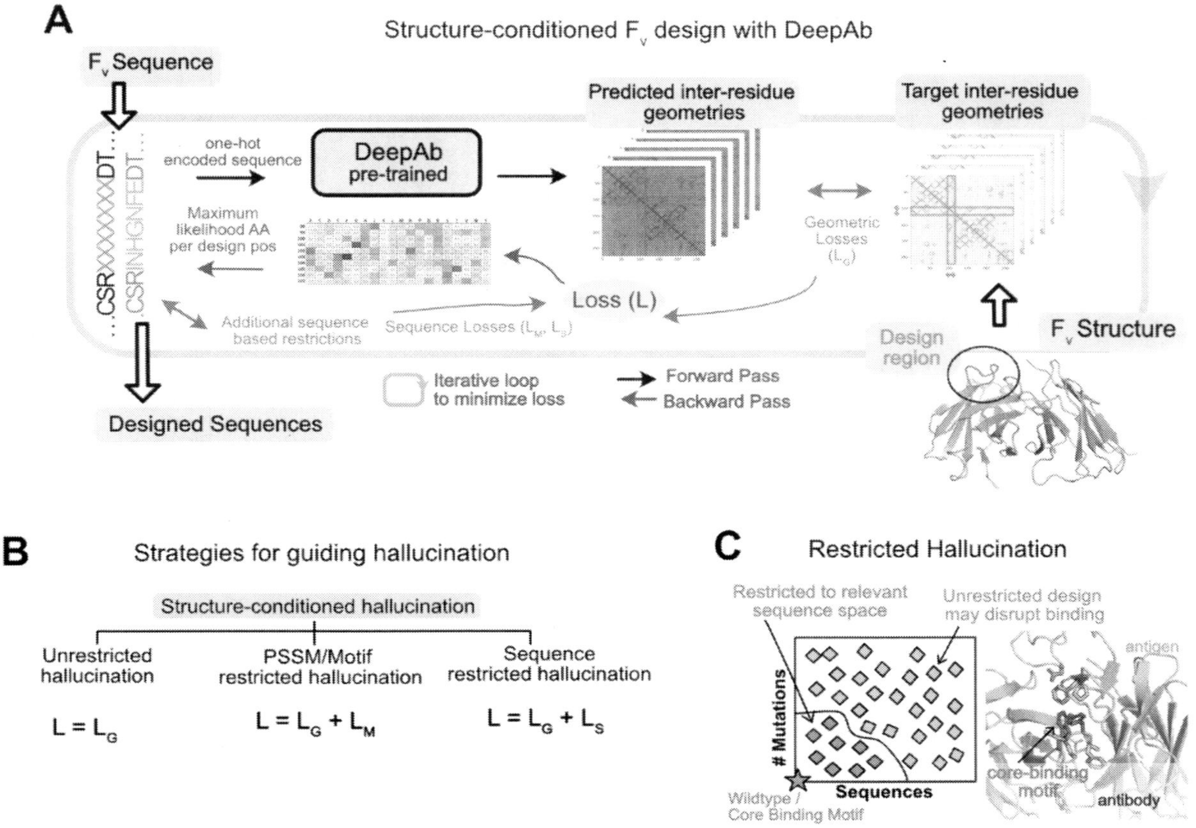

Figure 1. Framework for structure-conditioned hallucination of antibody subsequences

218i Identifying Novel Fentanyl Analogs from Mass Spectral Measurements

Arun S. Moorthy, Anthony J. Kearsley, William E. Wallace III

Illegal designer drugs are a rampant societal problem. Clandestine laboratories produce new drugs with molecular structures that are *sufficiently different* from currently scheduled drugs, which results in the drugs being more easily distributed before law enforcement can act. Fentanyl related compounds (FRCs) are one major class of designer drugs. We recently developed the NIST Fentanyl Classifier to accelerate the identification process of new FRCs. This method combines an evaluated mass spectral library of known fentanyl analogs, a new chemistry-informed feature similarity measure called *hybrid similarity*, and traditional clustering techniques to determine whether a compound is an FRC given its electron ionization mass spectrum, and produces a probable molecular structure if the compound is deemed sufficiently related to fentanyl. Using the NIST Fentanyl Classifier, we correctly propose structures for most FRCs that differed from fentanyl by the addition of one or two moieties. We incorrectly classified a limited set of "unrelated" compounds as FRCs, but these compounds only differed from fentanyl by the loss of a single chemical moiety. In this presentation, we cover the measurements and algorithms that underlie the NIST Fentanyl Classifier, focusing on how leveraging the predictable fragmentation of fentanyl for feature selection greatly simplified the problem. We will also discuss similar approaches and extensions to other chemical applications.

218j Developing a Defeatured Atom-Additive Model to Predict Single Component Partition Coefficients with FT-ICR MS Data

David Kenney, Heather LeClerc, Randy Paffenroth, Michael T. Timko, Andrew R Teixeira

Octanol-water partition coefficients (Kow) are a useful tool for determining solute-solvent partitioning behavior as it provides insights towards the lipophilic or hydrophilic nature of compounds. Single component values can be identified experimentally, from first principle ab initio thermodynamics or, more commonly, approximated by regressive algorithms. However, current methods fall short in understanding complex systems with thousands of unique compounds such as those present in oil spills and bio-oil production. Fourier Transformation Ion Cyclotron Resonance Mass Spectroscopy (FT-ICR MS) is a powerful analytical tool that identifies molecular formulas and relative abundances of ions based on their excitement in a magnetic field.

In this work, we developed a number of machine-learned models (linear regression, random forest, gradient boosted, etc.) that predict single component partition coefficients based on the data available through FT-ICR MS. By using web scraping methods, a database of 25,970 data points, with 5,514 unique molecular formulas, were collected along with their experimental partition coefficient value. The data was regressed using multiple techniques and found that partition coefficients could be determined on minimal information. Using an independent validation set of nearly 4,000 compounds, our model can produce a mean absolute error of 0.37. Combining this new regression algorithm with FT-ICR MS of complex oil-water systems provides insights into the molecular makeup and partitioning signatures of complex oils.

AUTHOR INDEX